DECENTRALISATION

Institutions, Justice and Social Development in India

DECENTRALISATION

Institutions, Justice and Social Development in India

P.N. SANKARAN
Chair Professor
Rural Management
St. Joseph's College of Business Administration
Bangalore

SERIALS PUBLICATIONS
NEW DELHI

First Published - 2006

ISBN: 81-8387-050-3

Published by

SERIALS PUBLICATIONS
4830/24, Prahlad Street, Ansari Road
Darya Ganj, New Delhi-110002 (India)
Phone : 23245225. Fax : 91-11-23272135
E-mail: serials@satyam.net.in

Dedicated to

MY LOVING PARENTS

&

COMMITTED TEACHERS

Whose words and deeds continues

To inspire my thoughts

THROUGHOUT

CONTENTS

FOREWORD

True to being a teacher, Prof. P.N. Sankaran has given a wider global perspective to the purpose and pursuit of decentralization-this also helps to understand better the value of the Kerala endeavour in decentralised governance and through it furthering justice and social development.

His scholarly labour has put in our hands a detailed treatise on decentralization–a vibrant contemporary theme of immediate relevance as also of long-term benefit to both State and Society.

The general propositions the book highlights are anchored on the real life experience of the State of Kerala in India which has acquired an instructive and inspiring place in Indian polity on development thinking and deft treatment of several national concerns especially demography, health and education.

Kerala's foray into 'decentralisation' is more recent than that of several other states. Though a late comer–Kerala today is in the lead ahead of other States in respect of both the substance and structuring of decentralization.

Rightly, Prof. Sankaran has devoted much thought and space to the Kerala design and its achievements. This indeed is the hard core of this book.

Decentralised planning and implementation are of crucial importance to polity in countries like India which have chosen democratic form of political governance. The hallmark of democratic governance is exacting 'accountability' of those who govern to the people at large, almost one to one and on a daily basis. This requisite of democratic governance came to be recognized by the political leadership in India only after it had gone through several general elections. Elected legislators–to parliament and state legislatures-increasingly faced embarrassment if not harassing questioning by the electorate: why they had not delivered the promised packages of development especially to the socially disadvantaged groups who suffer most from discrimination, deprivation and injustice and who otherwise command substantial voting power. The electorate also began to question more and more as to why they were not given any say in decision-making and in deployment of development resources which were mostly controlled by bureaucracy although it was well known that the bureaucracy had its ears open, if at all, only to top echelons and not to the people at the bottom.

This cascading ferment created enabling environment for decentralisation and exerted pressure on the political process to respond positively. In Kerala itself the doyen among political leaders–EMS Namboodiripad argued forcefully for taking 'democracy' right upto the grassroots. He also sought to end bureaucratic hegemony:

The Constitution itself according to me, failed to envisage an integrated administration in which, apart from the Centre and the States, there will be elected bodies, which will control the permanent services at the district and lower levels. Democracy at the Central and State levels, but bureaucracy at all lower levels - this is the essence of Indian polity as spelt out in the Constitution.

It was with such a centralized administration as its core that Panchayats were envisaged in the Constitution and the Balvantray Mehta Report. It is, therefore, not surprising that neither the bureaucrat nor the politician at the States level is prepared to decentralize whatever power has been conferred on the State under the Constitution. The point is to make a radical change in the very concept of democracy and adopt what is called four-pillar democracy.

Both from the political and socio-developmental angles, it is imperative to decentralize power, planning process and developmental activities below the state level.

All the staff engaged in development that is now available at the district and lower levels for these programmes will have to be placed under the elected representative bodies.

The centralized system of governance, which had dug in over the years refused to budge. However, at the state level, Kerala launched its own initiatives in decentralization within the constricted space.

Twenty years after Namboodiripad put forward his formulation on this theme, Rajiv Gandhi arrived on the scene as Prime Minister. During his travels to hundreds of villages he came face to face with the ugliness of over-centralized and under-decentralised system of governance in which people had not even a peripheral role in decision-making and planning and implementation of local development. He argued before Parliament for the liquidation of the existing unrepresentative and unresponsive system, for putting an end to it and replacing it with representative decentralized governance:

I toured hundreds of villages. I spoke to countless people. There, in their hearths and homes, I experienced the cruelty of an unresponsive administration, the oppression of an administration without a heart, the callous lack of compassion that most of our people find at the hands of much of our administration. At that time, I must confess, we were in quest of managerial solutions to an unresponsive administration. As we went along, we discovered that a managerial solution would not do.

We learnt that a grassroot administration without political authority was like meal without salt. We learnt that, however well-intentioned our district bureaucracy might be, without effective elected authority the gap between the people and the bureaucracy could not be closed. We learnt that the vacuum created by the absence of local-level political authority had spawned the power brokers who occupy the gap between the people and their representatives in distant assemblies and the even more remote Parliament. We learnt that corruption could only be ended by giving power to the panchayats and making panchayats responsible to the people. We learnt that callousness could only be ended by empowering the people to send their own representatives to institutions of local self-government… "the Panchayat Raj and Nagarpalika Bills are not only instruments for bringing democracy and devolution to every angan (front yard). They are also a charter for ending bureaucratic oppression, technocratic tyranny, crass inefficiency, bribery, nepotism, corruption and the million other malfeasance that afflict the poor of our villages, towns and cities. The bills were the warrant for ending the reign of the power brokers, of the intermediaries.

He persuaded Parliament to provide constitutional cover to decentralization to ensure that elected representatives 'ruled' and in turn were made to give an account of their promises and performance to the whole adult population every five years.

It is nobody's case, nor of Prof. Sankaran, that the situation has altered radically in this short span. But, the direction is right. This book shows that there is a visible and encouraging shift in the development landscape and massive activation of the people for self-government, which is the purpose and significance of decentralisation.

The students, teachers, elected representatives and practitioners of development and decentralization will all find rich fare served in this comprehensive volume for which we owe a debt to Prof. P. N. Sankaran.

—L.C. Jain

PREFACE

Development discourse, planning and practice have been undergoing rapid changes over the years, particularly in the last two decades. The strategy of decentralization, one among such issues for example, has taken different shapes in content and practice like de-concentration, delegation, devolution and privatization. Similarly, theorists speak of political, fiscal, institutional and administrative decentralization typologies also. Decentralization is the transfer of decision-making powers to more directly concerned lower levels of government and administrative authority. The interest in decentralization as a mechanism for transforming society is not new. However, a new wave of decentralization process is sweeping across different countries over the last two decades. National programmes on decentralization are generally presented as a means of streamlining national institutions and transferring political decision making power to the citizens and sometimes also as a key to wide- scale raising of living standards, better implementation of development projects and reduction of social disparities.

Decentralization, however, is an extremely broad concept in development strategy, covering a wide spectrum of general development policy goals. Its implementation involves a range of actors with varied and frequently conflicting interests. According to a recent report, democratic decentralization- truly giving voice to the people-requires much more than just decentralization and devolution of power. It also requires widening participation-especially by people who are often marginalized, such as women, minorities and the poor-and increasing the accountability of public officials/elected representatives at local levels, reforming/strengthening institutions, addressing issues in social sector development, human rights and decentralized dispensation of justice.

The present volume has been conceptualized in the above context of institution-building, inclusion of excluded communities/groups, participatory justice, accountability and transparency and social sector development under decentralization, which is viewed as a strategy that places people at the centre- stage of development as 'subjects' as well as 'objects' of development.

As in many other democracies, decentralization in India has been a process of advances and retreats, and a struggle between advocates of conflicting perspectives on governance and development strategy. However, a bold initiative by the Government of India paved the way for Constitutional sanction to the process of decentralization, set in motion by the 73rd and 74th Amendments to the Constitution. Kerala is one among the few states in India to seize the opportunity of decentralized governance and planning and contribute substantially to the practice of the new development paradigm. Despite the low growth rate and PCI, the state has been widely acknowledged as a unique model in social attainments such as high literacy rate, high life expectancy, low infant mortality rate and egalitarian social security measures. In the above backdrop, the experience of Kerala with decentralization over the last one decade would be of immense value to the academia, policy makers, development practitioners, LSGs and third sector institutions across the country and worldwide for continuation of the debate on decentralization and review/ dissemination of its design and practice.

The present volume is a compilation of research papers and evaluation studies published/ presented in academic and popular journals, books, seminars and conferences related to different dimensions of decentralization over the last ten years, authored by me with particular reference to Kerala, a pioneer in the decentralization experiment of learning by doing. I am grateful to the Kerala State Planning Board for my deputation to the Board (as Chief, Decentralized Planning Division in the launching phase of the Peoples' Planning Campaign for Decentralization) that paved the way for my interest in the dynamics of decentralization. Similarly, I would like to acknowledge the funding and other support rendered by the Kerala Research Programme on Local Level Development (KRPLLD), Centre for Development Studies (CDS), Trivandrum, Kerala, the Kerala State Planning Board and Institute of Rural Management Anand (IRMA), Gujarat and permission granted to publish the reports of the research projects, studies and assignments. I would like to acknowledge National Institute of Rural Development, Hyderabad, Institute of Social Sciences, New Delhi, Institute for Social and Economic Change, Bangalore and Publications Division, M/o I & B, GoI for bringing out the original publications which are included in the present volume. The valuable contributions of Dr.K.N.Nair (Director, CDS and Programme Coordinator, KRPLLD), Prof. P.R.Gopinathan Nair, (Senior Fellow CDS and Programme Advisor, KRPLLD), Dr.T.M. Thomas Issac (Professor, CDS and former Member, State Planning Board) and Dr. B. Vijayakumar (Reader in Economics, University College, Trivandrum and co-author of two research papers (4 and 8) included in the volume) in connection with the research work leading to this volume are thankfully acknowledged. Institutional facilities provided by University College, Trivandrum, Kerala, where most of the papers were developed, are gratefully remembered.

Fr. Dr. Aveline R D'Souza S.J, Director, St. Joseph's College of Business Administration, Bangalore deserve special gratitude for his uncompromising pursuit for excellence and value addition in research, that motivated me to bring out this volume. Shri L.C. Jain, former Member, Planning Commission and a living legend in the crusade for participatory democracy, has been kind enough to contribute a foreword to the volume. Thanks are also due to Ms Archana Ravindran (Research Associate), for research assistance, Mr. Leslie Fernandez, Ms. Geetha and Mr. Jijo Chacko (SJAIT) for necessary computer facility and support.

—P.N. Sankaran

Part - I

INTRODUCTION

1

DECENTRALIZATION: AN OVERVIEW OF GLOBAL AND REGIONAL SCENARIO

DECENTRALIZATION: AN INTRODUCTION

Decentralization is the transfer of decision making powers to more directly concerned, lower levels of government and administrative authority [Stock Mayer, 1999]. The interest in decentralization as a mechanism for transforming society is not new. In the second half of the twentieth century, practically every country has experimented with some form of decentralization or local government reform with varying aims and outcomes [Cheema and Rondinelli, 1983 review decentralization initiatives of the 1970s; Campbell et.al, 1991 review of recent Latin American experience]. There have, however, been a number of recent developments that distinguish the present wave of decentralization from earlier attempts. First, democratic institutions have been established and/or their role extended in many countries. Second, most of the countries presently involved in decentralization initiatives recognize the importance of providing financial resources to decentralized institutions to permit them to carry out their powers and responsibilities. Third, there is growing realization that many types of institutions can actively participate in decentralization efforts. It has been recognized that NGOs and community level organizations have a significant role to play in improving service delivery and for providing improved mechanisms for targeting disadvantaged groups [Parker, 1995]. These developments are likely to enlarge considerably the scope for overcoming some of the major factors that undermined earlier decentralization efforts, and to improve the prospects for sustaining decentralization initiatives once they have been established.

DECENTRALIZATION: THE OBJECTIVES

Generally the objectives of decentralization as put forward by the countries are:

- Promotion of a more balanced development in the country
- Design of more realistic projects and programmes which take into account local potentials and constraints

- More effective coordination of development activities at various spatial levels through disaggregation of planning functions
- Strengthening of local political institutions and increasing people's participation in development and
- Boosted mobilization of local resources [Maetz and Quieti, 1987].

DECENTRALIZATION: TYPOLOGIES

Rondinelli [1981] distinguishes between four different categories of decentralization:

- *De-concentration* is defined as a transfer of power to local administrative offices of the central government
- *Delegation* is the transfer of power to sub-national governments and/or para-statals, or other government entities
- *Devolution* is the transfer of power to sub-national political entities; and
- *Privatization* is the transfer of power to the private sector.

The "Training and Visit" [T&V] System of agricultural extension provides a good example of de-concentration of responsibility.[1] But its criticisms suggest that de-concentration is an incomplete strategy for decentralization. Between 1975 and 1987, the Brazilian government implemented 22 IRD projects in the ten states of Northeast Brazil. Both the Federal and State governments had responsibility for project organization. To provide a focus on the northeastern projects and to enable some decentralization of decision-making, the government established a parastatal[2] with overall responsibility for project planning, budgeting and M&E. Actual project implementation, however, was the responsibility of the states. Because of complexity in the approval systems and lack of integration into existing institutional structures, delegation also failed to deliver improved RD outcomes. A typical example of devolution relates to Panchayati Raj in India, where the Constitution required elections to the village and district levels of local governments. However, decisions concerning the assignment of functions and resource transfers to the local governments remained at the state level.[3] In response to criticism of poor service delivery from publicly owned institutions, and to fiscal crises in many developing countries, governments have embarked on programmes of privatizing services associated with RD.[4] However, there are concerns about the selective participation of the private sector, which necessitates a continued role for the public sector.

Apart from the Rondinelli terminology, a refinement in the dimensions of decentralization–political, fiscal and institutional-has become popular which explains how easy it is to undermine one type of decentralization by not giving sufficient attention to the other elements of decentralization.[5]

(a) Political Decentralization

Political decentralization involves transfer of policy and legislative powers from central governments to autonomous sub-national assemblies and local councils that have been democratically elected by their respective constituencies [Mc Lean 2002]. Benefits of political decentralization consist of greater voice and choice [of individual sub-national and local government constituents], better allocative efficiency and empowerment of people at the grass roots.[6] Decentralization initiatives that focus exclusively on electoral and administrative processes without considering the fiscal and institutional dimensions will not be sustainable (Parker, 1995).

(b) Fiscal Decentralization

In order to perform the responsibilities devolved to them, decentralized institutions need appropriate level of fiscal resources to cover the costs of providing public goods and services.[7] Sources of revenue available to them are own locally generated resources, transfers from higher level institutions and resources from borrowing.

(c) Institutional Decentralization

Although the decentralization experiments in the late 1980s and early 1990s did see the scope of institutions utilized, broadened to include elected local councils with devolved powers and responsibilities, other types of institutions like non-governmental organizations were still perceived as peripheral. According to Parker, concern with the institutional dimensions of development is a relatively new phenomenon. Earlier decentralization initiatives in Africa and Asia focused almost entirely on the administrative institutions of government. The institutional dimensions of decentralization are concerned with defining which formal government institutions are to be involved in a decentralization programme and the development of an appropriate legal framework governing inter-institutional relationships [Parker, 1995]. An active civil society[8] may assist significantly the implementation of decentralization.

Minis and Rondinelli (1989) propose a useful typology that distinguishes between three types of decentralization: spatial, market and administrative. Spatial decentralization, commonly called regional policy, is defined as a process of defusing urban population and activities geographically away from large agglomerations. Market decentralization, or economic liberalization is defined as a process of creating conditions in which goods and services are provided by market mechanisms rather than by government decisions. Administrative decentralization is defined as the "transfer of responsibility for planning, managing and the raising and allocation of resources from the central government and its agencies to field units of government agencies, subordinate units or levels of government, semi-autonomous public authorities or corporations, area-wide regional or functional authorities, or non governmental private or voluntary organizations" (Rondinelli and Nellis, 1986).

DECENTRALIZATION: NEW APPROACHES

Whereas the above approaches to decentralization are of conceptual interest and generally employed in public administration literature, a more detailed framework has been suggested by Maetz and Quieti [1987] to take care of the design requirements of training programmes in support of decentralization efforts. The framework is divided into four parts:

- Administration/Organization
- Finance
- Planning, and
- People's Participation, in line with what is considered as the four major preconditions of real decentralization.[9]

In response to structural and political drivers, East Asian countries have taken different approaches to decentralization, combining standard elements of delegation, deconcentration, and devolution found in many intergovernmental reforms around the world. Country approaches have been divided into three broad categories: *fast starters* (Philippines and Indonesia), *incrementalists* (China and Vietnam), and *cautious movers* (Combodia and Thailand) [White and Smoke, 2005].

DECENTRALIZATION IN SELECTED COUNTRIES

A brief outline of decentralization in selected countries and the motivations for decentralization are given below [Tables 1, 2, 3 & 4]. Degrees and Types of Decentralization by Region, Decentralization Worldwide, and Degrees and Types of Decentralization by Income, are summarized in Boxes 1, 2 and 3 respectively. It is noted from Box 1 that regions under Western Europe and the Americas take a lead in decentralization and that in all the regions the extent of political decentralization is the highest and it is followed by fiscal decentralization. Moving decision-making closer to people requires that citizens have voice and exit options for local governance (political decentralization). In addition, the local governments they elect, should be allowed home rule in fiscal, regulatory and administrative matters (fiscal and administrative decentralization). All of these elements must be in place to ensure effective decision-making at the local level. It is interesting to note that very few developing countries have adopted such a comprehensive approach to decentralization of decision-making. To take a stock of the progress of decentralization worldwide, for a selected group of countries representing different regional grouping, we take a comparative look at key aspects- political, fiscal and administrative decentralization.

(a) Political decentralization: This represents the area of most intensive political interest and reform. Following the lead of Latin America, other regions have also carried out reforms in support of political decentralization. Overall, the last two decades have witnessed major gains in political decentralization worldwide, with South Africa and Eastern Europe having completed their agenda for reform in this area. Other regions have also shown progress but progress has been quite slow in Central America and non-existent in the Middle East and North Africa regions.

(b) Administrative Decentralization: Effective administrative decentralization requires lack of any ex-ante controls over the decisions to hire, fire and set terms

Table 1
Political Decentralization in Selected Countries

	Constitutional safeguard against arbitrary dismissal of local govt.	*Popular election of local council members*	*Popular election of heads of local councils*	*Degree of popular participation in local elections*	*Provisions for popular recall of local officials*	*Contestability in local elections*	*Security of existence for local government*	*Overall political decentra-lization*
Africa								
Mozambique	No	Yes	—	Low	—	—	—	—
Nigeria	Yes	Yes	No?	Low	?	Low	?	Low
South Africa	Yes	Yes	No	High	Yes	High	Yes	High
Uganda	Yes	Yes	Yes	?	Yes	?	?	—
EastAsia/Pacific								
China	No	No	No	Low	No	Low	No	Low
Indonesia	Unclear	Yes	No	Medium	No	Low	No	Medium
Philippines	Yes	Yes	Yes	High	Yes	Medium	No	Medium
E. Europe/C. Asia								
Albania	Yes	Yes	Yes	Low	?	Low	No	—
Czech Rep.	Yes	Yes	Yes	High	?	High	No	High
Georgia	Yes	Yes	No	Low	?	Low	Yes	Low
Poland	Yes	Yes	Yes	Medium	Yes	Medium	Yes	High
Russia	Yes	Yes	Yes	Low	?	Medium/high?	Yes	Medium
Latin America								
Argentina	?	Yes	Yes	Medium	Yes	High	No	Medium
razil	Yes	Yes	Yes	Medium	Yes	High	Yes	High
Mexico	No	Yes	Yes	Medium	Yes	High	No	Medium
Mid. East/N. Africa								
Egypt	No	Yes	No	—	—	Low	—	Low
Jordan	Yes	Yes	Yes	Medium	—	Low/Medium	—	Medium
Morocco	Yes	Yes	No	—	—	High	—	Medium
S. Asia								
Bangladesh	Yes	Yes	Yes	Medium	Yes	High	No	Medium
India	Yes	Yes	No	Medium	Yes	Yes	Yes	High
Pakistan	No	Yes	No	Medium	Yes	Yes	No	Medium

Source: Shah and Thompson (2004), p. 6.

Table 2
Fiscal Decentralization in Selected Countries

	Rate & base setting for local revenues	*Majority of transfers are formula based & unconditional*	*Revenues more or less match responsibility*	*Own revenues finance majority of expenditure*	*Responsibility & control over all municipal services*	*Responsibility & control over health, education & welfare*
Africa						
Mozambique	—	—	—	—	Yes?	Yes
Nigeria	No	No	No	No	Some	No
East Asia/Pacific						
China	No	Yes	Yes	No	Yes	Yes
Indonesia	No	Yes	Varies by municipality	No	Yes	Yes
Philippines	—	—		—	—	—
E. Europe/C. Asia						
Albania	Rate only	Yes	No?	No	Yes	Yes
Czeh Rep.	No	Yes	No?	No	Yes	Yes
Georgia	Rate	Yes	No?	No	Yes	Yes
Poland	No	Yes	Yes	Yes	Yes	Yes
Russia	No	Yes	No	No	Yes	Yes
Latin America						
Argentina	Yes	Yes	Yes	Yes	Yes	No
Brazil	Yes	Yes	Yes	Yes	Yes	Yes
Mexico	No	Yes	Yes	No	No	No
Mid. East/N. Africa						
Eypt	No	No	No	No	No	No
Jordan	No	No	No	No	Yes	No
Morocco	No	Interpret data?	—	No	Yes	—
S. Asia						
Bangladesh	No	Yes	Yes	Yes	Yes	No
India	No	Yes	Yes	Yes	Yes	No
Pakistan	No	Yes	Yes	Yes	Yes	No

Contd....

Table 2
Fiscal Decentralization in Selected Countries

	Autonomy for planning	*Autonomy for procurement*	*Ability to borrow from domestic banks/higher levels of govt.*	*Ability to issue domestic bonds*	*Ability to borrow from foreign banks*	*Ability to issue foreign bonds*	*Overall fiscal decentralization*
Africa							
Mozambique	No	—	—	—	—	—	—
Nigeria	Shared	No	No	No	No	No	Low
South Africa	Yes	Yes	Yes	No	No	No	Medium
Uganda	Yes	?	Yes	No	No	No	Medium
East Asia/Pacific							
China	No	No	No	No	No	No	Low
Indonesia	Shared	No	Yes	Yes	Yes, through central govt	No	Medium
Philippines	—	—	—	—	—	—	—
E. Europe/C. Asia							
Albania	Yes	No	Yes	Yes	Yes	Yes	Medium
Czeah Rep.	Yes	No	Yes	Yes	Yes	Yes	—
Georgia	Yes	No	No	No	No	No	—
Poland	Yes	No	No	No	No	No	Low
Russia	No	No	No	No	No	No	Modest
Latin America							
Argentina	No	?	Yes	Yes	Yes	Yes	Medium
Brazil	Yes	Yes	Yes	Yes	Yes	Yes	High
Mexico	No	No	Yes	Yes	No	No	Low
Mid.East/N. Africa							
Eypt	Limited	—	Yes	—	—	—	Low
Jordan	—	—	Yes	Yes	—	—	Low
Morocco	Yes	—	Yes-from govt	—	—	—	Medium
S. Asia							
Bangladesh	No	No	Yes	Yes	No	No	Medium
India	No	No	Yes	Yes	No	No	Medium
Pakistan	No	No	Yes	Yes	No	No	Medium

Source: Shah and Thompson (2004), p. 7.

Table 3
Administrative Decentralization in Selected Countries

	Freedom to hire/fire/set terms of employment of local govt.	*Freedom to contract out own responsibiltes*	*Administrative regulatory authority (By- laws)*	*Overall administrative decentralization*
Africa				
Mozambique	—	—	—	—
Nigeria	No	No	Yes	Low
South Africa	No	Yes	Yes	Medium
Uganda	No	Yes	Yes	Low
EastAsia/Pacific				
China	No	Yes	No	Low
	De jure,yes;	De jure	Central govt.	Low
	De facto,	Yes; de facto	Approval	—
Indonesia	No	No	Required	—
Philippines	No	No	Yes	Low
E. Europe/C. Asia				
Albania	Yes	Yes	Yes	High
Czech. Rep.	Yes	Yes	Yes	High
Georgia	Yes	Yes	Yes	High
Poland	Yes	Yes	Yes	High
Russia	No	Yes	?	Low
Latin America				
Argentina	—	—	—	—
Brazil	Yes	Yes	Yes	High
Mexico	No	Yes	Yes	Low
Middle East/N. Africa				
Egypt	No	—	Yes	—
Jordan	—	—	—	—
Morocco	—	—	—	—
South Asia				
Bangladesh	No	Yes	Yes	Low
India	No	Yes	Yes	Low
Pakistan	No	Yes	Yes	Low

Source: Shah and Thompson (2004), p. 7.

of employment of local staff. To improve tax collection or the delivery of local public services, local government should have the freedom to contract own taxing and spending responsibilities. Furthermore, local governments should have the authority to pass bye-laws in their spheres of responsibility without having to obtain prior clearance from the higher-level government.

In practice, however, local governments, in a large majority of countries do not have the authority to hire and fire senior local staff. Eastern European countries represent an important exception in this regard. The freedom to contract own responsibilities is typically available but in some cases, e.g. in Philippines, this option is constrained to the expenditure side only. Regulatory authority for

Table 4
Motivations for Decentralization

Motivation	*Countries and/or Regions*
Political and economic transformation	Central and Eastern Europe, Russia
Political crisis due to ethnic conflict	Bosnia-Herzegovina, Ethiopia, Yugoslavia, Nigeria, Sri Lanka, South Africa, Philippines
Political crisis due to regional conflicts	Indonesia, Madagascar, Mali, Senegal, Uganda, Mexico, Philippines
Enhancing participation	Argentina, Brazil, Bolivia, Columbia, India, Pakistan, Philippines
Interest in EU accession	Czech Republic, Slovakia, Hungary, Poland
Political maneuvering	Peru, Pakistan
Fiscal crisis	Russia, Indonesia, Pakistan
Improving service delivery	Chile, Uganda, Cote D'Ivoire
To centralize	China, Turkey, European Union
Shifting deficits downwards	Eastern and Central Europe, Russia
Shifting responsibility for unpopular adjustment programs	Africa
Prevent return to autocracy	Latin America
Preservation of Communist rule	China
Globalisation and Information revolution	Most countries

Edited and abridged (2002), Courtesy: World Bank .

municipal services is usually available to local governments in most countries, although in Indonesia local government regulations have to be approved by the central government. Overall, there has been significant progress in administrative decentralization in developing countries in recent years. Such progress has been much slower in the area of local government autonomy for own civil service. In Pakistan, lack of such autonomy is considered the Achilles heel of devolution, as

Box 1
Degrees and Types of Decentralization by Region

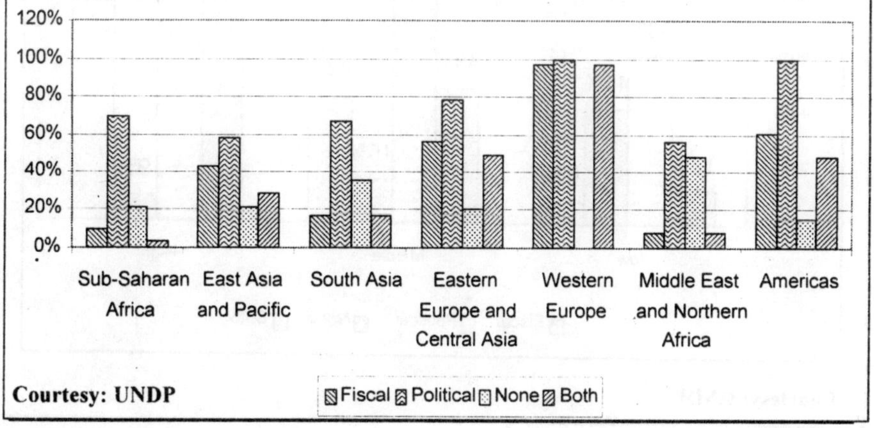

Courtesy: UNDP

Box 2

Decentralization Worldwide

- It is estimated that 80% of developing countries including the transitional economies of Eastern and Central Europe are experimenting with some form of decentralization.
- In 1999, 96 of 127 (76%) countries where politically decentralized. (At least one elected sub-national level of government).
- 42 or 33% had two or more elected sub-national tiers.
- 52 of the 127 countries (41%) had fiscal decentralization.

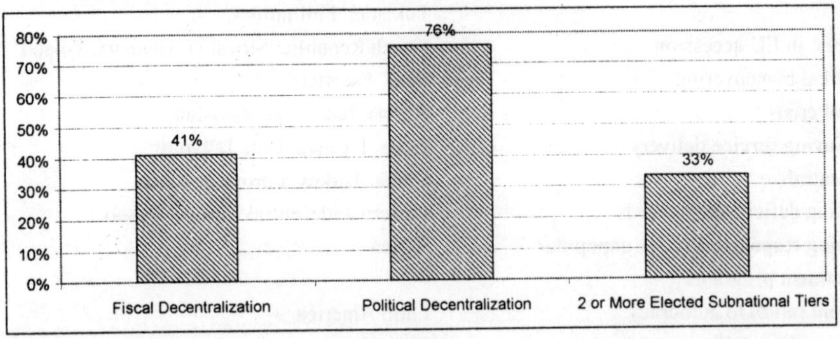

Courtesy: UNDP

Box 3

Degrees and Types of Decentralization by Income

- Degrees and types of decentralization vary by income.
- Countries with high incomes are likelier to adopt both fiscal and political decentralization.
- Countries with low GNP are least likely to devolve fiscal responsibilities to the sub-national level.

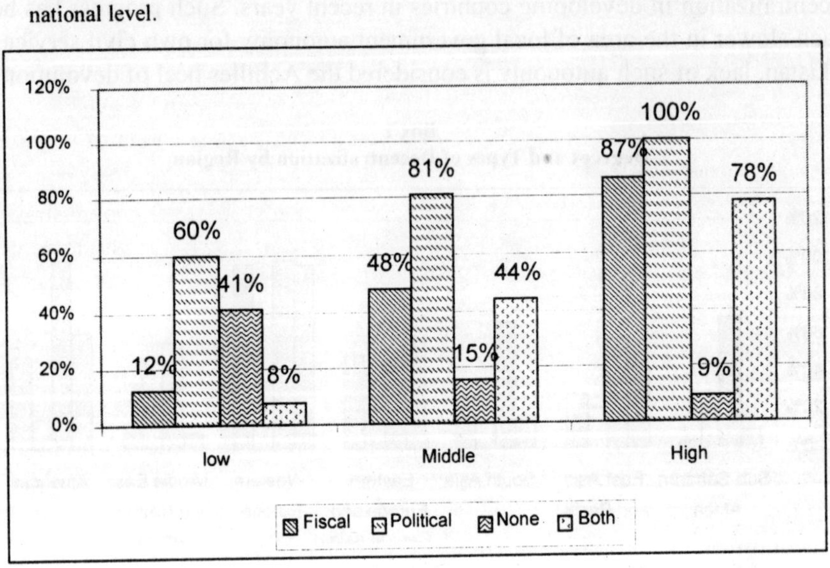

Courtesy: UNDP

the provincially transferred civil servants have no personal stake in the success of the devolution. In Indonesia, the central government sets the minimum salaries for civil servants of sub-national governments.

(c) Fiscal Decentralization: Fiscal decentralization has three important components: (a) revenue autonomy and adequacy, (b) expenditure autonomy and (c) borrowing privileges.

Since decentralization is a multi-dimensional process fraught with success and setbacks, the related initiatives will have to be subjected to a continuous process of modification reflecting changes in social, political and economic conditions. Cheema and Rondinelli [1983] recognized the importance of the political, fiscal and institutional dimensions of decentralization, but none of the authors explicitly related these three dimensions to RD outcomes [Parker,1995]. In the above context Parker has developed a conceptual model for analyzing decentralization called the *soufflé* theory of decentralization[10], which provides a useful framework for analyzing the separate dimensions of decentralization and their inter-linkages and permits the investigation of patterns of decentralization across countries.

A wave of new decentralization process is sweeping across different countries, over the last one and a half decade. The reasons governments give for decentralization programmes are not always the same as donors' grounds for supporting them. National programmes are generally presented as a means of streamlining national institutions and transferring political decision making power to the citizens and sometimes also as the key to wide-scale raising of living standards and reducing social disparities. Donors justify support to national programmes because:

- Decentralization puts relations between the citizen and the administration and between local and central government on a market economy footing;
- Decentralization is a response to the changed functions and regulatory capacity of the state and the more differentiated demands of the public purse;
- In some cases decentralization is also justified on the grounds of economic restructuring [Stockmayer, 1999].

Thus decentralization is an extremely broad concept in development strategy, covering a wide spectrum of general development policy goals. Its implementation involves a range of actors with varied and frequently conflicting interests. Decentralization processes are largely uneven, seldom one-way, extremely complex and subject to diverse policy influences [Beier, Christoph 1999]. In a similar tone Prudhomme [1994] observes that decentralization measures are like some potent medicines: they must be taken at right time, in the right dose, and for the right illness to have the desired salutary effect. While examining the dangers, or costs associated with decentralization it is noted that the benefits of decentralization have probably received more attention than its costs. According to the author:

- The benefits of decentralization in allocative efficiency are not as obvious as suggested by the standard theory of fiscal federalism

- These doubtful benefits might carry a cost in terms of production efficiency
- Decentralization runs counter to redistribution and stabilization
- Decentralization makes re-distributive policies, whether interpersonal or inter-jurisdictional, more difficult, if not impossible, and
- Decentralization also makes macroeconomic stabilization programmes more difficult to implement because sub-national government fiscal policies can run counter to national policies.

These are serious drawbacks or potential drawbacks, which should be taken into consideration in the design of any decentralization programme [Prudhomme, 1994]. United Nations Development Programme [UNDP] in the *Human Development Report 2002* 'Deepening Democracy in a Fragmenting World' deals with democratic decentralization in a number of countries including Bolivia, United Kingdom, Vietnam, China and India.[11] According to the Report, in principle, decentralization of power from the center to provinces, district or villages enables people to participate in decision-making more directly. But in reality it can simply transfer power from one set of elites to another. Democratic decentralization-truly giving voice to the people–requires more than just decentralization and devolving power. It also requires widening participation-especially by people who are often marginalized, such as women, minorities and the poor-and increasing the accountability of public officials at local levels. The Report is critical about India's Panchayati Raj, which is reported to be prone to capture by elites and to subversion by central political authority. The worldwide trend towards decentralization has been accompanied by animated discussions about its benefits and costs........Whatever the reason, there are signs that some countries are now ready to move forward and the next decade may see as much implementation as rhetoric. India may be a case in point [The World Bank, 2004].

DECENTRALIZATION IN INDIA: EVOLUTION AND CURRENT STATUS

In India, one of the largest democracies in the world, a federal union of 28 states inhabited by a population of over a billion people, the most significant experiment in reforming governance from a participatory democratic decentralization point of view has been the introduction of the constitutionally mandated Panchayati Raj system through the 73rd Constitutional Amendment in 1992.

As in many other democracies, decentralization in India has been a process of advances and retreats, and a struggle between advocates of conflicting perspectives on governance and development strategy [The World Bank, 2004]. The political advocates among the supporters have argued that decentralization is essential for appropriate participation in government; technical experts have supported decentralization as critical to the effective implementation of development programmes. The opponents have included politicians who see decentralization as diluting political control by putting power into ill-equipped hands; and administrators, who perceive parallel threats to public management. At every step, there have been delays in legalization and effectiveness measures. Despite the two Constitutional Amendments (73rd and 74th) and affirmative legislation by states that underpin decentralization, the overall picture, according to the World Bank, remains an unfinished agenda: a patch-work of legislation and initiatives that has put local governments in place but failed to endow them with the means to deliver results.

DECENTRALIZATION: A HISTORICAL PERSPECTIVE

Decentralization to rural and urban governments (Panchayats and Municipalities) began even before independence. Self-governing village communities characterized by agrarian economies existed in India during the Rig-Vedic period. In course of time such village bodies metamorphosed into *'panchayats'*.[12] The panchayats have had a chequered history in the country.[13] With the advent of the British rule in India, and centralized administration, the institution went into decay in most parts of the country (Wadia, 1996). Decentralization efforts under British Raj are listed *in* Box 4.

Despite Mahatma Gandhi's commitment to Gram Swaraj or Village Self-government, the subject of decentralization was included only in Part IV of the Constitution dealing with Directive Principles of State Policy.[14] The Panchayati Raj movement in India was placed in the national agenda in 1957 with the constitution of the Belwant Rai Mehta Committee to recommend measures for Community Development and National Extension Service Programmes (See Box 5).

Box 4
Decentralization: Key Facts under British Raj

1687	British formed the fist Municipal Corporation in Madras.
1870	Mayo Resolution to bring about administrative efficiency.
1870	Revival of gram Panchayats through Bengal Chowkidari Act.
1880	Famine commission highlighted inefficient disbursement of relief measures due to absence of local bodies
1882	Ripon Resolution elevated the role of local administration.
1906	Indian National Congress adopted self–government as the political goal for the country under the Presidentship of Dadabhai Naoroji.
1909	Government constituted the Royal Commission on Decentralization.
1909	Lahore Resolution of INC-"to make all local bodies from village Panchayats upwards elective with elected non–official Chairmen".
1919	Montague& Chelmsford Reforms-bringing the local bodies under the domain of Indian ministers in the provinces.
1925	Eight provinces in British India adopted the reforms-Madras, Bombay, Bengal, Central Provinces & Berar, UP, Punjab and Assam.
1926	Six native states, Cochin, Travancore, Indore, Baroda, Kolhapur, Mysore adopted village panchayat laws.
1935	Government of India Act allowed provincial autonomy, facilitating the evolution of Grama Panchayats.

Source: Kaushik, 2005.

The Committee recommended the organization of a three-tier structure of the PRIs for effective decentralization with elected panchayats, finance, power and

Box 5
The Balwant Rai Mehta Committee's Recommendations

- Constitution of statutory elected local bodies with the necessary resources, power and authority.
- Decentralized administrative system working under the control of statutory bodies.
- Basic unit of democratic decentralization should be located at the block/ samiti level.
- Envisaged elected bodies 'panchayats' for a village or group of villages
- An executive body called Panchayat Samiti for a block, with directly elected and co-opted members.
- An advisory body called Zilla Parishad at the district level
- Chairman as the district chief, elected through ex-officio members from the lower tier and others.

Source: Kaushik, 2005.

authority supportive of Community Development projects. Andhra Pradesh and other states followed the PRI structure, inaugurated in Rajasthan in 1959. K. Santhanam Committee was set up in 1963 to look into panchayat finance, for which the B.R. Mehta Committee did not make provisions. The Santhanam Committee proposed a special tax on land revenues and homes and the establishment of a Panchayat Raj Finance Corporation to look into the financial resources of PRIs and provide loans/assistance. However, the PR movement failed to take roots since there was a deliberate attempt by the bureaucracy, vested interest groups and elected representatives to thwart regular elections because their ascendancy was feared (Mathew, 2001).

The PR system was left in hibernation until the Janatha government, inspired by the ideals of Jayaprakash Narain constituted the Ashok Mehta Committee to review the system. It proposed a two-tier model of decentralization with Zilla Parishad and Mandal Panchayat. In order to ensure participation of weaker/ marginalized groups (women, SC & STs), the Committee proposed reservation of seats and posts for them. In terms of its recommendation to include panchayats in the Constitution, the Committee paved the way for the transition of panchayat as a development orgnisation to a political organization [Box 6]. The GVK Rao Committee 1985 and the L.N. Singhvi Committee (1986) further reviewed the PRI system.[15]

DECENTRALIZATION: CONSTITUTIONAL MANDATE

All the committees in their reports listed three prominent reasons for the failure of PR system in achieving the objectives of decentralization:

- States did not devolve enough substantive power to the PRIs;
- Bureaucracy did not cooperate with PRIs, and
- The traditional local elite groups manipulated the PRIs in their own self interest.

Box 6
The Ashok Mehta Committee's Recommendations

- Favoured two-tier model for panchayati raj–zila parishad & mandal panchayat.
- Direct elections to these bodies, reservation of constituencies for SC/STs and women.
- State governments should not supercede the PRIs, all development functions should be put under the zila parishad.
- Preparation of exhaustive list of functions based on location–specific programmes.
- PRIs should be allowed to mobilize resources by granting them powers of taxation.
- Improved interface between PRIs, voluntary agencies, cooperatives and local government institutions for capacity building and human resource development.

Source: Kaushik, 2005.

Such assessments made the government to think in terms of giving the Panchayats a constitutional status. Subsequently the 64th Amendment Bill (1989) introduced in the Parliament suggested setting up of three tier panchayats systems at village, intermediate and district levels; the Bill culminated into the 73rd Amendment to the Constitution, which was passed in both Houses of Parliament in December 1992.

The amendment was ratified by most of the states and came into effect from April 24, 1993.[16] The 73rd amendment was inserted in part IX of the Constitution [Box 7].

The states and UTs amended their respective PR Acts by the deadline April 1994 to pave the way for three tier local self-governments (Table 5). States with a population below 20,00,000 were given the option of not having the intermediate level panchayat.

POLITICAL SPACE TO THE INDIGENOUS PEOPLE-PESA 1996

As the provisions of the 73rd constitutional amendment did not apply to Scheduled Areas, located in eight states (namely AP, MP, Rajasthan, Gujarat, Maharashtra, H P, Orissa and Bihar), the Parliament extended the Act to these areas by legislating the Panchayats (Extension to the Scheduled Areas) Act 1996. The basic features of PESA are listed in Box 8.

PESA gives radical self-governance powers to the tribal community and recognizes its traditional community rights over natural resources (especially minor forest produce), ensuring the customary mode of conflict resolution etc.

The Act placed the grama sabha at the center of the PR system of self- rule for the *Adivasis*, going beyond the functions and powers of the grama sabha in non-scheduled areas.

Thus the 73rd Amendment covers many areas that would enable the panchayats to perk up the lives and well being of poor, vulnerable and marginalized groups. It also makes specific provisions that promise the participation of traditionally excluded groups, such as women, SCs, STs, and transparency for local institutions. Local bodies are expected to move away from their traditional role of simply

Box 7

73rd Constitutional Amendment: Basic Features

- Constitutional status to the panchayats and gives it uniformity by making the three–tier system a permanent feature.
- Panchayats to be constituted in every State at the village, intermediate and district level, except States with a population not exceeding 20 lakh.
- Establishing a Gram Sabha in each village, exercising powers and performing functions at the village level, as the State may provide by law.
- Members of panchayats at all levels will be elected through direct elections. The election of the chairperson at the intermediate and district level will be through indirect elections.
- Seats are reserved for SCs and STs at all levels according to their population. Not less than one-third of seats are reserved for women.
- The office of Chairperson will also be subject to this provision.
- A uniform five-year term has been granted to the panchayats. In case of dissolution, elections should be held within six months of the date of dissolution.
- Panchayats are entrusted with (i) preparing plans and implementing schemes for social justice and economic development; (ii) in regard to matters listed in the 11th Schedule.
- State legislature authorizes the panchayats to levy, collect and appropriate suitable local taxes. The Government can make grant-in- aid to the panchayats from the Consolidated Fund of the concerned State.
- Review of the financial position of the panchayats will be undertaken by a finance commission, which shall be constituted every five years.
- State Election Commission shall be constituted to ensure free and fair elections to the panchayats.

Source: Kaushik, 2005.

Box 8

PESA Basic Features

- Every village will have a grama sabha that will be competent to safeguard and preserve tradition and customs of the locals and their cultural identity.
- Prior approval from the grama sabha/gram panchayat is mandatory before granting permission by the government.
- Grama Sabha/panchayat were given ownership of minor forest produce.
- Grama sabha/ panchayat were endowed with powers to manage village markets.
- Grama sabha/ panchayat were endowed with poweres to lend money to tribals

Source: Kaushik, 2005.

executing the programmes handed down to them by higher levels of government and to formulate and implement their own programmes of economic development and social justice (Jha, 2000). For the first time, panchayats with a high degree of uniformity in terms of structure, composition, powers, and functions have taken shape throughout the country. There are approximately 600 District Panchayats, 6000 Blocks or Taluka Panchayats and 2,50,000 village panchayats, covering 72.2 per cent of India's population. Almost all states have held at least one round of

Table 5
State-wise Conformity Acts with the 73rd Amendment Act

Sl. No.	*State/UT*	*Year*	*Name of the Act*
1.	Andhra Pradesh	1993	The Andhra Pradesh Panchayati Raj Act 1993
2.	Bihar	1993	The Bihar Panchayati Raj Act 1993
3.	Gujarat	1993	The Gujarat Panchayati Raj Act 1993
4.	Himachal Pradesh	1994	The Himachal Pradesh Panchayati Raj Act 1994
5.	Karnataka	1993	The Karnataka Panchayati Raj Act 1993
6.	Kerala	1994	The Kerala Panchayati Raj Act 1994
7.	Madhya Pradesh	1994	The Madhaya Pradesh Panchayati Raj Act 1994
8.	Maharashtra	1994	The Bombay Village Panchayats & the Maharashtra Zilla Parishad & Panchayat Samitis (Amendment) Act, 1994
9.	North East		
	(a) Assam	1994	The Assam Panchayat Act 1994
	(b) Tripura	1993	The Tripura Panchayat Act 1993
	(c) Manipur	1994	The Manipur Panchayat Act 1994
	(d) Arunachal Pradesh	1994	The Arunachal Pradesh Panchayati Raj Ordinance 1994
10.	Orissa	1993	The Orissa Gram Panchayat, Taluk Panchayat, Zilla Parishad (Amendment) Act, 1993
11.	Punjab	1994	The Punjab Panchayati Raj Act 1994
12.	Rajasthan	1994	The Rajasthan Panchayati Raj Act 1994
13.	Tamil Nadu	1994	The Tamil Nadu Panchayati Raj Act 1994
14.	Uttar Pradesh	1994	The Uttar Pradesh Panchayat (Amendment) Act 1994
15.	West Bengal	1994	The West Bengal Panchayat (Amendment) Act 1994

Source: Task Force Status Report, 2000

elections since 1993 in accordance with the provisions of the Act. Voter participation in those elections has also been exceptionally high (Alsop et al. 2000).

DECENTRALIZATION IN INDIA: KEY CONCERNS

- The conformity legislation (Principal Legislation) allowed state governments to prescribe new rules without going through legislative processes (to deal with new contingencies). This permits the state to withdraw, curtail, extend, and override the authority of the PRI system. States such as AP and Karnataka have used the route of delegated legislation in preference to principal legislation (World Bank, 2004).
- Such flexibility given to the states resulted in wide variation in the political structure of Panchayats across India (size, method for electing chairpersons, reservation for Backward Classes [BCs], election norms, procedure for dissolution, suspension and removal of members/governments etc.) (World Bank, 2004).
- In many states fiscal decentralization has been incomplete and inadequate. While states such as Kerala have devolved substantial resources to Panchayats, others such as Uttar Pradesh have managed only minimal devolution. Moreover, there is no common approach followed by all the State Finance Commissions

[SFCs] especially in assessing the resource requirement. Instead of assigning tax from the state to local bodies, many SFCs favour devolution of a higher percentage of the state's shared taxes (World Bank, 2004).

- As a result of ambiguity in devolution of functions (coming under 29 subjects) many activities and sub-activities continue to be implemented by line departments (GOI, 2000).
- Because of weak endowment of administrative powers to LSGs, government administration in India continues to be centralized, despite being highly decentralized politically.
- A recent review of ten years of Panchayat Raj in Madhya Pradesh, Tamil Nadu and Kerala by Narayana (2005) indicated evidence of elite capture of LSGs[17] and that extremely poor, landless and those living in thatched houses have not become members to the extent of their share in total population. Based on similar findings, it is concluded that the evolution of Panchayats as LSGs has been slow and tardy.
- It is one thing to pass a law but quite another to make it work. The passing of laws has not led to local self governance owing to three sets of factors- lack of devolution of powers and resources, lack of capacity building and poor involvement of women (Narayana, 2005).
- There is also considerable evidence of human rights violations in the process of decentralization, involving Dalits, BCs, Tribals and Women.[18]

DECENTRALIZATION IN KERALA: AN OVERVIEW

Contemporary development literature has acknowledged Kerala's unique social attainments, such as high literacy rate, high life expectancy, low infant mortality rate and egalitarian social security measures-all achieved despite a low growth rate and a low rate of industrialization. Kerala has now added a new dimension to its achievements with the state's remarkable progress towards decentralized planning and participatory governance following the 73rd and 74th Constitutional Amendments [Oommen, 2004]. As a prelude to the overview of the decentralization process in Kerala, the state's prior experience in the field is described first.

(a) The Early period

Viceroy Lord Rippon's resolution of 1882 establishing local bodies called "*little republics*" in urban centers [later extended to rural areas] was the first decentralization initiative in the history of erstwhile Kerala, carved out from the Princely States of Travancore and Cochin, and the Malabar district of the former British Indian Province of Madras Presidency. Later, following the recommendations of the Royal Commission on Decentralization [1909], a two tier local government system was introduced in Malabar-the Malabar District Board at the district level and the Panchyat Board at the village level. The formation of the Panchayat Acts in Cochin and Travancore and traditional village governments in the region were also influenced by the Government of India Act 1919. Later in 1925, a Village Panchayat Act was passed in Travancore and seven Panchayats were formed. A good number of important functions and taxes were assigned to them along with a system of grants.[19] In the Malabar district, village panchayats functioned under the Madras Village Panchayats Act [1950]. Their main sources of revenue were a share of land cess and duty on the transfer of property. House tax, profession tax and vehicle tax were also levied. As the panchayats were not eligible for any statutory grants, the government supported them through ex-gratia grants. Areas not covered by panchayats were governed by the administrative set up of the Malabar District Board and they derived their income from profession tax, entertainment tax, education cess, surcharge on stamp duty, license fees and government grants-in-aid.

(b) Later Period

From the first Kerala ministry [1957] to date, the history of Kerala's decentralization process has been influenced by the ups and downs of coalition politics in the state [World Bank, 2004]. When the B.R Mehta Committee was at work, Government of Kerala constituted a committee for suggesting administrative reforms on August 15, 1957 headed by the Chief Minister. The terms of reference of the committee, inter alia, covered suggestion of methods for democratization of the organs of government of various levels with a view to effective participation of local self governing institutions, and other representative bodies in the administration. Following the recommendations of the Committee [July 26, 1958], Kerala Panchayat Bill and Kerala District Councils Bill were introduced on December 9, 1958 and on April 16, 1959 respectively. But none of these could be enacted into law. Subsequently, the Kerala Panchayat Act 1960 and the Kerala Municipal Corporation Act 1961 were enacted unifying the existing laws of Travancore-Cochin and Malabar regions and enlarging the functions and financial resources of the local bodies. In February 1964 the Kerala Panchayat Union Councils and Zilla Parishad Bill were introduced in the Assembly on the line of the B.R. Mehta Committee recommendations. Block was conceived as the basic unit for planning and development. The Kerala Panchayat Raj Bill 1967, which contemplated a two tier system: Panchayat at the basic level and Zilla Parishads at the district level also could not become law. Subsequently the Kerala District Administration Bill 1971 introduced in the Assembly also met with the same fate. However, the same bill was introduced in the Assembly in 1979.[20]

Following the broad outline suggested by the Report of the Planning Commission Working Group on District Planning, the first phase of initiation was started in Kerala by decentralizing to the district level the formulation and implementation of Special Component Plan [SCP] and Tribal Sub Plan [TSP][21] in 1983-84 [GOK, State Planning Board 1991]. Encouraged by the results of the decentralization process covering SCP and TSP, the government decided to extend decentralization to other areas as well; as the next phase of decentralization it was decided to formulate District Plans as part of Eighth Five Year Plan exercise.[22] As part of the exercise, District Development Councils [DDCs] were constituted with Sub-committees for agriculture and allied sectors, industry and infrastructure and social services. Under this model, plan formulation exercise stared from the panchayats and moved upwards through the blocks to the district. Decentralization process in Kerala received further impetus in 1988 with the acceptance of the key recommendations of the Special Advisor to government (V. Ramachandran), appointed for suggesting amendments and administrative arrangements to the Kerala District Administration Act.[23] The new District Council system definitely represented a bold step towards democratic decentralization. The Steering Committee on Decentralized Planning and Planning for Rural Development, Eighth Five Year Plan (1990-95) recommended further steps towards strengthening decentralization; it recommended acceptance of V. Ramachandran's recommendations on the powers, functions and responsibilities of District Councils. In the opinion of the Steering Committee, Panchayats were to be provided with an apparatus to adjudicate upon aspects standing detrimental to development activities in the form of *'Nyaya Panchayats'* with powers to dispose of certain civil and criminal cases.[24]

(c) The Modern Period

Though Kerala attempted to usher in a new era of democratic decentralization and Panchayati Raj in the first four decades of its formation, it was the Constitutional Amendments 73rd and 74th which brought a new life to the decentralization drive in the state. The Kerala Legislative Assembly adopted the Panchayat- Nagarapalika Bills in 1994; the first elections to the three-tier PR setup were held in September 1995 and the LSGs came into being on October 2, 1995. The government launched the *People's Plan Campaign* [PPC] on 17 August 1996.[25] The Committee on Decentralisation of Powers (1996) laid down the principles and laws of local governments, which were incorporated into the Kerala Panchayati Raj Act. What distinguishes Kerala from the other states in India is the 'big-bang' approach the state has taken in decentralization. The strategy was to implement a significant fiscal decentralization programme and then "learn by doing". Indeed, it was a question of transferring powers, then setting up procedures and precedence for their exercise [World Bank 2004]. Decentralization programme of Kerala encompasses 991 villages or Gram Panchayats [GPs], 152 intermediate level Panchayats [Block Panchyats or BPs] and 14 District Panchayats [DPs].

OBJECTIVES OF DECENTRALIZATION/PEOPLE'S PLAN CAMPAIGN

In a broad sense, the campaign was an attempt to conceptualize, operationalise and institutionalize a system of multilevel, people–centered planning process suitable to the regional specificities of Kerala which would also help overcome contemporary development crisis (State Planning Board, Kerala (nd). The People's Campaign for Ninth Plan represents an initiative to make use of the legacy of collective social intervention and the strength of mass movements to meet the contemporary crisis of development (Issac and Harilal, 1997).

The specific objectives of the campaign consisted of:

- Reconstructing the received ideas on decentralization and people' s planning and enabling the state government to conscientise and mobilize people for development action
- Imparting knowledge and skills needed for local level planning and involving mass organizations of people in development activities, and
- Inculcating a favorable attitude among officials and technocrats so as to facilitate people's articulation of development needs, projects, implementation strategies, monitoring and integration of plans of LSGs, vertically and horizontally.

PHASES IN PEOPLE'S PLAN CAMPAIGN

Despite the fact that on many occasions, the campaign had to reorient its programme, the process of decentralized planning consisisted of the following concrete phases:

- **Phase 1** (August-October 1996): convening of Grama Sabha and Ward Committees for identifying local development problems, supported by training of (key) resource persons at state, district and local levels.
- **Phase 2** (October-December1996): organization of Local Development Seminars and drafting of Development Reports, describing Grama Sabha reports, review of ongoing schemes, collection of secondary data, geographical study of the area and a brief survey of local history.
- **Phase 3** (December-January 1997): constitution of about 12 sector- wise Task Forces to projectise the recommendations of Development Seminars. The task forces that comprised of elected representatives (Chairpersons) and officers of line departments (Conveners) were to draft reports stating objectives, beneficiaries, activities, organizations (for implementation), financial and cost-benefit analysis and monitoring mechanism.
- **Phase 4** (February-April-1997): prioritization and incorporation of projects into five year plans of LSGs. Each panchayat prepared a Plan Document structured into introduction, development strategy, resource mobilization, sectoral programs, integrated development, welfare of SC/STs and women, and monitoring. A team of Voluntary Technical Corps (VTC) undertook preliminary vetting of the plans.
- **Phase 5** (March-April 1997): preparation of Block and District Plans (own and integrated).
- **Phase 6** (April–1997): holding Grama Sabha/ Ward Conventions to discuss the plan document and select beneficiaries.

The above exercises were followed by year-end and mid-term reviews of progress and corrections wherever required.

STEPS IN DECENTRALIZATION

The following steps have been significant in the decentralization exercise:

- Legal entitlements of local governments
- Transfer of institutions and staff and demarcation of functional domain [Health, Education, Poverty alleviation, Social welfare, Agriculture, Local roads, Sanitation and water supply, tiny, cottage and small industries, Welfare Pensions etc.] and
- Transfer of financial resources [more than one third of Plan resources and rural local governments getting about 85 per cent in proportion to the rural population and 15 per cent going to urban governments].

The unique features of financial devolution consists of:

- Around 90 per cent of the Plan funds [untied] to be distributed towards: productive sectors [40%], infrastructure [30%] and gender sensitive schemes [10%] for Women Component Plan
- The entire plan grant is investible [pure money] devoid of staff salary or administrative costs
- All plan grants to LSGs are separately budgeted [Annexure-IV of State Budget.]
- Village Panchayats obtain 70 per cent of grants and 30 per cent goes to District and Block Panchayats
- Transparent formula is followed in financial devolution [based on population, area, own income, composit index of agricultural labourers and backwardness]
- Practice of operationalising decentralization, through a participatory planning methodology, and
- Towards institutionalization[26] [from experimentation phase through a corrective phase and later the critical institutionalization phase].

Towards genuine decentralization, Kerala has gone considerably ahead of other states in terms of instituting and strengthening the following institutions:

- The State Election Commission
- The Finance Commission [First SFC 1994, Second SFC 1999 and the Third SFC 2004]
- Ombudsman for local governments
- Appellate Tribunals [authority at regional and district levels to take care of appeals by citizens against decisions of LSGs], and
- State Development Council [headed by the Chief Minister, members of the Cabinet, Leader of Opposition, Chief Secretary, Vice-Chairman State Planning Board, Presidents of District Panchayats and representatives of other tiers of LSGs].

PLAN INTEGRATION AND DISTRICT PLANNING COMMITTEE

Preparation of district plan is a constitutionally mandated responsibility of the District Planning Committee [DPC]. Article 234 ZD [1] lays down that "there shall be constituted in every state at the district level a District Planning Committee to consolidate the plans prepared by the Panchayats and Municipalities in the district and to prepare a draft development plan for the district as a whole". The DPC is visualized only as a coordinating agency and the whole responsibility is to integrate the local plans prepared by the LSGIs. A close reading of the constitutional provisions also makes it abundantly clear that the functional role of the district plan is to ensure the mechanism for integration of the multi-level plans within the district. Despite having the unique constitutionally accorded status, only in a few states have any serious attempts been made to empower the DPCs[27] and only in rare instances have the district development plans been actually drawn up. One of the most important events in 1999-2000 has been the formulation of district plans by the DPCs.

In Kerala, the district plan (DP) is conceived not merely as a consolidated list of projects and plans of different tiers of LSGIs but as a comprehensive document providing a macro perspective for the development of the district [GOK, 1999]. Since the district planning process has been put in place, local plans at every level would be prepared with simultaneous feedback from both above and below. The functions of the DP has been defined as follows:

- To make an objective assessment of the district and an assessment of the problems of resources use and provide a macro perspective for sustainable development of the district
- To analyse the consolidated plan of the Local Self Governments in the district so as to identify contradictions, duplications and gaps and also link the local plans with the macro perspective, and
- To formulate guidelines for the preparation of future local plans so that the weaknesses identified can be avoided in future and the local plans are better integrated with the state level plan.

Accordingly DP in the Kerala context has three parts:

Part-1: Perspective for district development consisting of an assessment of the resource base, development problems of productive, services and infrastructure sectors, separate analysis of the problems faced by women and weaker sections, all with due emphasis on the spatial dimensions of physical planning.

Part-2: Consolidated plans of the LSGs with relevant information with respect to the pattern of plan finance, sectoral investment, Special Component Plan [SCP], Tribal Sub Plan [TSP], and Women Component Plan [WCP] at different tiers. The consolidated plans of the local bodies are juxtaposed with the State and Central plan schemes proposed for the district.

Part-3: Guidelines for preparation of future local plans are formulated so as to take care of the related issues:

- To what extent do the local plans correspond to the district macro development perspective of Part-1?.
- What are the reasons for the divergence and how could they be rectified?.

- To what extent do the local plan proposals correspond to the local problems identified?.
- To what extent are the local plans integrated and consistent with each other?.

Part 2 and 3 of the DP are prepared every year, whereas Part 1 is prepared only once in five years at the beginning of the five year plan. The DP document is then discussed in detail at the State government level and the conclusions of the discussions are presented at the State Development Council.

SCENARIO ON THE EVE OF A DECADE

Among the different states in India, Kerala is accepted as the leader in the post 73rd and 74th Constitution Amendment decentralization experiment. In the last ten years, it was able to structure a decentralized system of development administration, following classical fiscal decentralization principles with emphasis on untied funding, clear demarcation of functions on the principles of subsidiarity, structuring of a transitional local governance system, setting up of independent umpiring institutions, facilitating people's participation and above all backing of all these with powerful laws [GOK, 2005].

Documentation of the experience of Kerala in decentralization over the decade is yet to be undertaken. Nevertheless, the following review of the performance of LSGs during 2003-04 brings out meaningful reflections on the trends and sustainability of decentralization in the state. Table 6 indicates that the performance of LSGs, in terms of the percentage of funds released from the budget provision (93.64) and the percentage of expenditure from the funds released (88.70), has been good. Expenditure pattern of LSGs during 2003-04 (Table 7) with nearly half the amount (48.64%) devoted to the service sector and the least emphasis given to the productive sector (18.91%) implies that their (LSGs) funds utilization is not fundamentally different from the broad pattern in the state and that LSGs in general have been neglecting the productive sector compared to the infrastructure sector (32.45%), which is better off relative to the status in other states. Given the unique rural-urban scenario in Kerala (sometimes characterized as 'rurban' where the rural-urban divide is less sharp), the level of expenditure (67.65%) by Grama Panchayats (Table 8) is a reflection of the commitment of the state to develop villages and the participatory forces that sustain the rural governments. Sector-wise expenditure of LSGs (Table 9) indicates little difference compared to Table 7 as far as the relative neglect of the productive sector is concerned. A look at the physical achievements of LSGs in different sectors during 2003-04 (Table 10) as an elaboration of the financial achievements brought out already, endorse the attraction of LSGs towards infrastructure and service sectors. Analysis of the sources of funds of LSGs (Table 11) during 2003-04 reveals that their dependence on government funds (grant-in-aid 72.88 % and centrally sponsored scheme 10.08 %) comes to nearly 83 per cent , their own funds constitute only 5.40 per cent and voluntary plus beneficiary contribution amounts to 4.63 per cent of total funds. However, in the case of grama panchayats the share due to centrally sponsored scheme is low (5.56%) and the contribution from beneficiaries is slightly high (5.53 %) compared to LSGs as a whole. Without a comparison against other states, budget provision per LSG (Table 12) does not convey much meaning; however, given the larger population in LSG s, and the allocation of 30-35 per cent of the state plan funds to them, the volume of funds transferred should be comfortable for their developmental activities. A comparison of Table 9 with 13 holds an indication that they tend to spend more on infrastructure (30.33% as against the allocation of 20.76%), which should not be the case in Kerala where infrastructure is comparatively developed. The LSGs received relatively higher allocation for productive sector (22.31%) and service sector (51.76%) against actual expenditure of 17. 68 per cent and 45.47 per cent respectively for the two sectors; part of the funds has been diverted to infrastructure development that has higher visibility in physical terms. It is stated that District Panchayats and Block Panchayats have set apart lower allocations for the productive sector (19.96 % and 20.32 %) compared to Village Panchayats (25. 31%), though the same mandatory sectoral provisions bind them all. Similarly, Municipal Corporations spend far less on the productive sector than Municipalities (GOK, 2005).

Some of the other issues identified, necessitating corrective action are:

- Persisting departmentalism-even after a decade, organic and harmonious link between transferred institutions and LSGs have not been generally established.
- Inadequacy of critical support systems in the formulation and vetting of local government projects.
- Visible setback to participatory systems in technical sanction for projects and bureaucratization of the related procedures.
- Participation in Grama Sabha and social audit practices are less enthusiastic.
- Accountability and transparency mechanisms continue to be problematic.

Table 6

Performance of Local Governments in Kerala 2003-04 [Rs. crore]

Sl. No.	*Category*	*Budget Provision*	*Total Grant in Aid released*	*Expenditure*	*Percentage of Release to Budget provision*	*Percentage of Expenditure on Release*
(1)	(2)	(3)	(4)	(5)	(6)	(7)
1.	General	923.92	889.44	816.94	96.27	91.85
2.	SCP	263.3 3	225.78	182.41	85.74	80.79
3.	TSP	45.31	32.66	22.61	72.08	69.23
4.	EFC	79.55	80.79	67.90	101.56	84.05
	Total	1312.11	1228.67	1089.86	93.64	88.70

Source: Government of Kerala, Economic Review 2004.

Table 7

Expenditure Pattern of Local Governments in Kerala 2003-04 [Rs. crore]

Sl. No	*Category*	*Productive Sector*	*Service Sector*	*Infrastructure Sector*	*Total*	*Un-classified*	*Grand Total*
(1)	(2)	(3)	(4)	(5)	(6)	(7)	(8)
1.	General	174.34	305.67	272.12	752.13	64.81	816.94
2.	SCP	11.79	125.02	44.80	181.61	0.79	182.40
3.	TSP	1.53	18.60	2.08	22.21	0.41	22.62
4.	EFC	4.99	46.26	11.53	62.78	5.12	67.90
	Total	192.65 (18.91)	495.55 (48.64)	330.53 (32.45)	1018.73 (100.00)	71.13	1089.86

Source: Government of Kerala, Economic Review 2004.
Figures in bracket indicate percentage to total.

Table 8

Expenditure Pattern of LSGs in Kerala 2003-04 [Rs. crore]

Sl. No	*Tier of Local Governments*	*General*	*SCP*	*TSP*	*EFC*	*Total*
(1)	(2)	(3)	(4)	(5)	(6)	(7)
1.	Grama Panchayats	542.47	122.32	15.08	57.46	737.33
2.	Block Panchayats	11.42	35.97	5.16	2.02	154.57
3.	District Panchayats	42.10	8.40	2.24	0.00	52.74
4.	Municipalities	95.45	12.68	0.13	7.35	115.61
5.	Corporations	25.50	3.03	0.00	1.07	29.60
	Total	816.94	182.40	22.61	67.90	1089.85

Source: Government of Kerala, Economic Review 2004.

Table 9

Sector-wise Expenditure of LSGs in Kerala 2003-04 [Rs. crore]

Sl. No	*Tier of Local Governments*	*Sector*				
		Productive	*Service*	*Infrastructure*	*Exempted*	*Total*
(1)	(2)	(3)	(4)	(5)	(6)	(7)
1.	Grama Panchayats	144.35 (19.58)	338.94 (45.97)	204.20 (27.69)	49.84 (6.76)	737.33 (100.00)
2.	Block Panchayat	27.34 (17.69)	69.73 (45.11)	42.41 (27.44)	15.09 (9.76)	154.57 (100.00)
3.	District Panchayat	12.44 (23.59)	26.77 (50.76)	13.22 (25.07)	0.31 (0.59)	52.74 (100.00)
4.	Municipalities	7.74 (6.69)	46.43 (40.16)	56.24 (48.65)	5.20 (4.50)	115.61 (100.00)
5.	Corporations	0.77 (2.60)	13.69 (46.25)	14.45 (48.82)	0.69 (2.33)	29.60 (100.00)
	Total	192.64 (17.68)	495.56 (45.47)	330.52 (30.33)	71.13 (6.53)	1089.85 (100.00)

Source: Government of Kerala, Economic Review 2004.
Figures in bracket indicate percentage to total.

Table 10
Selected Physical Achievements of LSGs Kerala 2003-04 (provisional)

Sl. No.	*Item*	*Unit*	*Total*
(1)	(2)	(3)	(4)
1.	Area brought under cultivation	Ha	58401
2.	Beneficiaries of vegetable cultivation	No	211939
3.	Distribution of tractors under agricultural development projects	No	67
4.	Distribution of tillers under agricultural development projects	No	506
5.	Poultry –egg rearing units	No	32504
6.	Poultry-broiler unit	No	1237
7.	Construction of cattle sheds	No	15943
8.	Watershed development projects	No	1553
9.	Land brought under cultivation through watershed development projects	Ha	11730
10.	Micro enterprises started	No	6276
11.	Self employment units started	No	5521
12.	Self employed persons	No	24901
13.	Self employment units started by poor people	No	4055
14.	Self employment units for destitutes	No	467
15.	Employment training programmes for destitutes	No	394
16.	Destitutes trained (persons)	No	462
17.	Distribution of land to landless [beneficiaries]	No	15320
18.	Area of land distributed to landless	No	2662
19.	Houses constructed	No	50426
20.	Construction of sanitation units	No	174689
21.	Drinking water projects	No	11926
22.	Beneficiaries of drinking water projects	No	307489
23.	Drinking water projects implemented with beneficiary contribution	No	3035
24.	New wells	No	21033
25.	Renovation of drinking water projects	Sq.m	14706
26.	Mobility assistance devices, Crutches, Walkers	No	2265
27.	Wheel chair, Tricycle, mechanized three wheeler	No	2223
28.	Hearing aids	No	1771
29.	Braille kit, study materials	No	757

Source: Government of Kerala, Economic Review 2004.

- Effort in local resource mobilization has been very weak and the fiscal crisis of the state is getting transmitted to local governments through problems in funds flow [GOK, 2005].

NEW INITIATIVES

In order to strengthen decentralization, the following initiatives have been taken:

- Implementation of the recommendations of the Second State Finance Commission
- Constitution of the Third State Finance Commission
- Activity Mapping- demarcating the areas of responsibility of different tiers of local government to achieve greater clarity on the functional responsibility of LSGs

Table 11
Sources of Funds and Expenditure of LSGs in Kerala 2003-04

(Rs. crore)

Tier of LSG	*Grant in aid*	*Own fund*	*State sponsored scheme*	*Centrally sponsored scheme*	*Externally aided Fund*	*Loan from cooperative Institutions*	*Loan from Financial Institutions*	*Voluntary contributions*	*Beneficiary contribution (direct)*	*Others*	*Total*
{1}	(2)	(3)	(4)	(5)	(6)	(7)	(8)	(9)	(10)	(11)	(12)
Grama Panchayats	737.33	62.86	18.08	55.04	6.80	0.95	3.44	5.37	54.73	44.87	989.47
BlockPanchayats	157.57	0.20	5.45	79.27	1.19	0.93	6.84	0.01	7.57	1.23	260.26
DistrictPanchayats	52.74	0.35	9.37	6.63	0.00	0.00	0.00	0.00	0.02	1.56	70.67
Municipalities	115.61	11.54	2.97	10.26	0.00	0.11	0.62	0.22	1.49	0.61	143.43
Corporations	29.60	6.00	0.00	0.00	0.00	0.00	0.00	0.00	0.00	0.01	35.61
Total	1092.85	80.95	35.87	151.20	7.99	1.99	10.90	5.60	63.81	48.28	1499.44
	(72.88)	(5.40)	(2.39)	(10.08)	(0.53)	(0.13)	(0.74)	(0.37)	(4.26)	(3.22)	(100.00)

Source: Government of Kerala, Economic Review 2004.
Figures in bracket indicate percentage to total.

Table 12
Budget Provision to Local Governments in Kerala 2004-05

(Rs. lakhs)

Sl. No.	Local Governments	No. of LSGs	Budget Provision			Amount per Local Government		
			Plan	Non-Plan	Total	Plan	Non-Plan	Total
1.	Grama Panchayats	991	77837.54	23745.63	101583.07	78.54	23.96	102.51
2.	Block Panchayat	152	22006.39	901.36	22907.75	144.78	5.93	150.71
3.	District Panchayat	14	19749.90	1525.59	21275.49	1410.71	108.97	1519.68
4.	Municipalities	53	15131.11	2427.17	17558.28	285.49	45.80	331.29
5.	Corporations	5	11693.31	879.24	12572.55	2338.66	175.85	2514.51
	Total	1215	146418.25	29478.89	175897.14	120:51	24.26	144.77

Source: Government of Kerala, Economic Review 2004.

Table 13
Sector-wise Plan Grant to LSGs in Kerala 2004-05

[Provisional] (Rs crore)

Sl. No	Local Governments	No. of Projects	Sector-wise Plan Grant				
			Productive	Service	Infrastructure	Others	Total
1.	Grama Panchayat	117715	187.73	376.51	133.44	43.93	741.60
2.	Block Panchayat	13679	31.62	83.38	31.06	9.56	155.62
3.	District Panchayat	2741	20.56	55.90	25.57	0.96	102.99
4.	Municipalities	8044	11.59	55.01	32.88	4.18	103.65
5.	Corporations	705	2.04	17.34	12.99	0.00	32.37
	Total	142884	253.54	588.14	235.94	58.63	1136.24
	% Allocation		22. 31	51.76	20.76	5.16	100.00

Source: Government of Kerala, Economic Review 2004.

- Setting up of the Appellate Tribunal[28] to serve both as accountability-enforcing mechanism against illegal and biased decisions in the regulatory sphere of LSGs
- Service Delivery Projects of MGP have been taken up in 104 Village Panchayats, 14 Municipalities and 5 Municipal Corporations, to improve LSG services in tune with Citizens' Charter Initiative[29], and
- Decentralization Support Programme [DSP], with the help of Royal Netherlands Embassy, has commenced with particular focus on capacity building.

DECENTRALIZATION IN KERALA: SOME OBSERVATIONS

In the light of the historical outline and recent experiences of Kerala in decentralization, the following observations merit immediate attention towards sustaining the process and corrective action. In terms of a comprehensive view of plan integration (at spatial, sectoral, financial, goal-based, rural–urban, inter-tier, and organizational levels, a recent study found that there has been limited success in plan integration in Trichur district, Kerala (Sudhakaran, 2004). In a micro level study of two GPs it was noted that most of the programs were more traditional in nature and were marked by unequal people's participation (Thomas, 2001). Similarly, a panchayat level evaluation study concluded that resource mobilization from sources other than plan funds has been unsatisfactory in all sectors (Mohanan, 2001). Another district level evaluation of decentralization confirmed that three fourth of the women respondents had actively participated in grama sabha, followed by task force meetings and plan preparations (Giridharan, 2001). An evaluation of people's planning in Thiruvananthapuram district revealed that the projects addressed artisans as a whole only in few panchayats and that traditional artisans were not covered at all, though they are the worst victims of modernization, technology change, shortage of critical raw materials, fluctuations in income and employment. For an effective coverage of artisans under decentralization programme, the study suggested an artisan plan component in selected artisan clusters of the state (Sankaran, 2001). In a recent UNDP study focusing on the contribution of decentralization on poverty eradication in Kerala, Prof. M.A. Oommen noted that, while the overall outcome definitely helped to deepen participatory democracy and poverty reduction, there are several negative aspects that need to be considered to put the wheels of democratic decentralization and development on a stable track. While transparency guarantees, right to information, citizens' charter, accountability to grama sabha and so on are accepted principles of decentralized governance in Kerala, they are neither institutionalized nor operationalised except in very exceptional cases. The decentralization episode of Kerala, however, has been held highly successful for the demonstrated synergy between Kudumbasree (the state poverty eradication mission) and the LSGs (Oommen, 2004). Similarly it is noted in another study that despite their obvious theoretical advantages, performance and social audit mechanisms may not prevail over other parallel systems of audit: the former is participatory while the latter is official. In the present environment of ambiguity, the latter tends to dominate the former [Sankaran, 2005].

In a recent study it is observed that there has been conspicuous decline in participation as indicated particularly by the lower voter turnout in the recent (2000 as well as 2005) local body elections viz-a-viz elections to the state assembly or parliament (Harilal, 2005). Another study noted that the plan performance of the grama panchayats has been witnessing deterioration and that the emphasis of plan process shifts from development to distribution of benefits (Prakash, 2005). According to Pillai (2005) the one single factor that accounts most for not fulfilling completely the objectives set forth in Kerala's attempt in decentralized governance and development has been the failure in adequate capacity building. Though Kerala leads among other states in decentralization, there is a need to have a re-look at its fiscal decentralization process, which is crucial and an important complement to its functional and administrative initiatives (Mathew, Anand 2005). In a recent paper Mukhopadhyay (2005) observe that the Campaign for democratic decentralization in Kerala was no doubt sagacious,participative and vigorously implemented. However, there was no audit certification to vouch for expenditire incurred or to counter allegations of "decentralized corruption". His contentions are based on two sets of assessments:

- Whether the Campaign was carried out within the framework of the provisions of the LSG Acts, verifiable objectives, accounting systems etc, and
- How far it empowered citizens and institutions?

Though the achievements of LSGs and the larger public in terms of capacity building and participation in decentralization are widely shared and well documented, issues of compliance with the provisions of the PR Act and accountability concerns are not addressed effectively.[30] According Mukhopadhyay, social audit as communicative action—drawing the experience of the Mazdoor Kisan Shakti Sangathan (MKSS) in the context of *jan sunwais* [public hearing] in Rajasthan—can help foreground the pretext of accountability and unleash the energies of people to establish accountability in local bodies of Kerala. Another suggestion is to increase the number of members of Ombudsman and requiring them to pass judgments only at the Jan Sunvais.

However, it is beyond doubt that Kerala has marched ahead in the road towards sustainable decentralization, developed best practices in local level planning (Box 9) and institution-building and extended planning process to address more effectively issues of women and other weaker sections of the society. The experience of Kerala with District Planning Committee, documentation of local history, resource mapping, capacity building, promotion of leadership through PRIs and SHGs, innovative systems in auditing and reporting, participatory practices in plan formulation, execution, monitoring and evaluation etc has valuable lessons for self- evaluation (learning by doing) and for replication by other states.

Box 9

Best Practices in Local Governments, Kerala

Nedumbassery Grama Panchayat

Resource mobilization of Rs. 1.62 crores from different sources which accounted for about 56% of the total Plan outlay of Rs. 2.92 crores (which included grant-in–aid, own fund, central & state sponsored schemes). *Destitute Plan:* Costing Rs.75 lakhs under implementation, identification of house sites [19] & construction of houses for the poor [24], organization of micro enterprises.

Panmana Village Panchayat

Efficient collection of tax & other revenue with more than 100% increase in 20 items during the last 4 years (building, service & profession tax, shopping complex, slaughter house, license fee, pensions, deposits, receipts from various departments etc.

Changanassery Municipality

Recycling of solid waste & Manufacturing of Farm Yard Manure by employing Vermi Compost method with project components to improve dumping yard infrastructure & recycling of solid waste at a cost of Rs 3.56 lakhs. Kudumbasree (State urban poverty alleviation project) engaged in collection & transportation of waste to the manure manufacturing unit.

Maneed Village Panchayat

Group Farming introduced in 16 paddy fields covering an area of 300 ha to improve productivity of HYV variety '*Jyothi*'. The effort was supplemented by investment in mechanization& lease-land-farming.

Elikulam Grama Panchayat

Fluctuation in the price of the major cash crop (rubber) was met with *Vanilla Development Project* under extension support of the Spices Board in 2002-03 covering 201 farmers. Local *Krishi Bhavan* supplied saplings & provided 1/3 subsidy of the total cost of cultivation. Processing unit on the anvil.

Vadakarapathy Grama Panchayat

Launching of Potteri *Community Irrigation Scheme* to solve water scarcity [where average rainfall is 60% less than the state average [by linking the Chinnaeri pond with the Potteri pond through a channel of 400 m, benefiting a command area of 50 acres. The work was undertaken by Beneficiary Groups.

Edayur Grama Panchayat

Rain Water Harvesting under a World Bank supported project '*jalanidhi*' (Kerala Rural Water Supply & Environmental Sanitation Project) undertaken by Women Development Groups with training in construction of ferro-cement tanks for water harvesting. Helped people lacking access to water & latrine.

Kollam District and Pulinkunnu Grama Panchayat

District–wide *Participatory Spatial Planning* by DPC initiated with technical support from Town & Country Planning Department, involving 15 year perspective & a 5 year execution plan. *The steps* are: collection & computerization of data, analysis, joint identification of priority areas for development, planning principles & strategies, vetting of analysis, development seminar, approval of draft plan by DPC, discussion with LSGs &finalization.

Pathanapuram Grama Panchayat

Waste to Energy Plant using waste from meat & fish market established in 2003, under subsidy of the M/o Non-conventional Energy, GOI. Gas generated from 250 kg of waste, in the treatment plant is used for cooking, lighting, running diesel engine and for generation of electricity.

Paravur Municipality

Vermi Composting Unit using solid waste [70 tons per day] was set up with assistance from *Peerumedu Development Society*, an NGO experienced in vermi composting. The current production rate is 300 kgs of compost & the Municipality is earning about Rs. 1.2 lakhs per annum.

Venganoor Grama Panchayat

Special School for Physically & Mentally differently abled Children under assistance from Kudumbasree Mission- Services of trained teachers, child psychologists & physiotherapists are made available-Students & the school supported by a fund with contribution from Panchayat grant-in-aid, Social Welfare Department, GOI & public.

10 Village Panchayats from Wayanad District

Panchayat, NGO- Government Partnership ' *The Giridhara Water Supply & Sanitation Project* for Tribals, using Central Assistance executed by an NGO *Socio Economic Unit Foundation* with involvement of Panchayat & Beneficiary Groups called *Ooruvikasana Samitis*. Unique in terms of beneficiary contribution by Tribals (eg. for 12 schemes, 218 families raised Rs. 4,40,000).

Mangattidam Grama Panchayat

Series of Poverty reduction projects (poor people identified in terms of 9 risk factors) like construction of houses, latrines, promotion of employment through 83 micro enterprises benefiting 730 families and a host of other projects in agriculture, animal husbandry, school education etc . As a result 44% of the poor families have crossed the poverty line.

Pilicode Village Panchayat

Winner of 'Nirmal Grama Puraskar' The village with 5103 families selected as one among the 38 Village Panchayats in the country for outstanding excellence in Environmental Sanitation. A winner of '*Grama Swaraj*' Trophy in 1997-98, the Panchayat had started environmental sanitation in1996. Annual Plan 2004-05, aimed at providing toilets to the remaining 249 houses.

Source: Developed from, Government of Kerala, Economic Review 2004.

NOTES

1. Developed during the 1960s, the T&V System aimed to address criticisms of existing extension arrangements. In India, the T&V System was introduced state by state, replacing the system of multi-purpose village level workers.
2. Known as SUDENE (the North-east Regional Development Agency), it was not well integrated into the pre-existing institutional structures and therefore became irrelevant and often a hindrance to project implementation.
3. Most states in India starved these local governments of resources and let the electoral process at the village level and district level breakdown.
4. With regard to agricultural extension, for example, there are now many instances where the private sector has become involved in service provision that was previously managed by the public sector (agro-processing firms, input suppliers, farmers, associations, media companies and consulting firms).
5. For example the PR experiments in India failed to decentralize fiscal resources to local institutions effectively, rendering useless the attempt to devolve political power on its own. In Brazil, failure to consider existing institutional arrangement left SUDENE isolated and irrelevant to the decentralization process (see Parker 1995).
6. Citizens have greater powers to influence decisions, which affect their lives and the local governments can meaningfully respond to constituency concerns.
7. Development Papers No.11 UNESCAP 1991, that draws heavily from a seminar held by ESCAP in collaboration with NIPFP New Delhi (1989) to review the experience of selected countries in the ESCAP region in the matter of decentralization, observes that the motivation for such a focus was that fiscal decentralization constitutes the core of the decentralization process in the economic sense, underpinning it financially and providing it with real content. For a detailed account of Fiscal Decentralization and Mobilization of National Resources, see UNESCAP, 1991.
8. 'Social capital' has been suggested by Putnam (1993) as a useful concept in the dynamics of institutional decentralization.
9. Four major preconditions of real decentralization are:
   - Government commitment to decentralization and some deconcentration
   - Some financial autonomy of the sub-national levels both in resource mobilization and decision making on its allocation
   - Clear demarcation and delegation of planning responsibilities to the lower levels, and
   - Existence of a local political backing to sub- national planning, without which sub-national proposals have very few chances of standing the arbitration stage and of being translated into action (Maetz and Quieti, 1987, p. 6).
10. Like a souffle that requires the right combination of milk, eggs and heat to rise, a successful programme of decentralization will need to include just the right combination of political, fiscal and institutional elements.
11. According to the Report, one of the most interesting development in decentralization over the past decade has been the expansion of people's participation and public officials' accountability at the local level in China and Vietnam.
12. A local body consisting of five village elders.

13. After fits and starts of establishing them during the British period, they were set up in most parts of the country after independence. The PRIs were on their ascendancy during 1959-64, stagnated during 1965-69 and entered a phase of decline thereafter. The 73rd Constitutional Amendment Bill passed in 1993 visualises the transfer of powers and ensuring elections being held regularly. The Bill also gave authority to panchayats to undertake development programmes, provided for grants-in-aid from the Consolidated Fund of India for the purpose, and setting up SFCs every five years, for providing funds to the Panchayats. (For a detailed historical perspective, See Wadia, F.K 1996).
14. According to Article 40, "the state should take steps to organize village panchayats and endow them with such power and authority as may be necessary to enable them to function as units of self-government", which was not mandatory.
15. The Rao Committee endorsed district as the basic unit of planning and programme implementation, while the Singhvi Committee strongly recommended creation of viable panchayats and devolution of financial resources to PRIs.
16. When both houses of Parliament passed the 72nd (Panchayats) and 73rd (Nagarapalika) Constitutional Amendment Bills of 1991, the sequence of the Bills changed to the 73rd and 74th Amendments respectively.
17. A number of other studies also show that like earlier attempts at decentralization, the 73rd Amendment has also not fully lived upto its expectations.
18. For a detailed review of human rights violations in different states, see Sahay (2005). Also see Mathew, [2003] for an account of human rights, panchayat institutions interface and the argument that the local governments based on vibrant democracy at the local government level with the support of civil society organization will protect the human rights of the people at all levels.
19. Construction of wells and tanks, communication, primary education, minor irrigation, public health, sanitation, lighting of public ways and places, collection of land cess, profession tax, vehicle tax etc.
20. The Bill received the President's assent on May 18, 1980.
21. Working Groups were constituted in districts, with the Collector as Chairman, District Planning Officer as Convener and the Additional Development Commissioner [ADC] and designated officers of line department as members. These nominated members were responsible for the implementation of SCP and TSP programmes coming under their respective sectors. Nearly 12 per cent of the Plan outlay was earmarked for SCP and TSP. See, GOK, State Planning Board 1991, for a detailed account of budgeting and other procedures.
22. As part of the exercise, plan schemes were classified into State Sector Schemes and District Sector Schemes. The criteria for allocation of outlays to the Panchayats and Municipalities/Municipal Corporations consisted of:
    - 50 per cent on the basis of population
    - 15 per cent on the basis of agricultural workers
    - percentage outlay equivalent to the percentage of SC/ST population
    - 10 per cent of the outlay equally among the Blocks
    - 2 per cent of the outlay among Municipalities and Municipal Corporations, and
    - the balance outlay for district level schemes.

23. Based on the recommendations of the Advisor, necessary amendments were made in the District Administration Act 1979 and the first elections to the District Councils were held in January 1991. These institutions started operation on the 5th of February, 1991.
24. The Committee recommended '*Nyaya Panchayats*' consisting of 5-7 members elected from the Gram Sabha , also representing women and SC/ST Groups in the panchayat area.
25. In fact the People's Planning Campaign [PPC] was the culmination of a long record of tentative initiatives and experiments in this area on the part of several agencies including the government. For example, the Panchayat Level Development Planning [PLDP] project taken up by the *Kerala Shastra Sahithya Parishat* [KSSP], a popular people's science movement, in 1995-96, with financial assistance from the Kerala Research Programme on Local Level Development [KRPLLD, carried out by the Centre for Development Studies, Trivandrum Kerala, as part of the Multi-annual Multidisciplinary Research Programme Initiative of the Netherlands Ministry of Development Cooperation] was the direct progenitor of the PPC. For a detailed account of the approach and activities of PLDP, see 'Panchayat Level Development Planning: Problems and Prospects of People's Participation in a Party-dominated Democracy', 6th Annual Report 2001-02, KRPLLD, CDS, Trivandrum. Another example is Panchayat Level Resource Mapping, or People's Resource Mapping Programme [PRM], as it was later called, undertaken by the Centre for Earth Science Studies [CESS] Trivandrum, Kerala in collaboration with other research and development organizations [Kerala State Land Use Board-KSLUB, Integrated Rural Technology Centre-IRTC, Department of Science and Technology-DST,GOI, Science, Technology and Environment Committee-STEC, GOK and State Planning Board].The PRM programme was conceived as a process of integrating scientific evaluation of each panchayat on its land and water resources, on the one hand, and the mapping of existing land use and assets by local volunteers on the other, leading to a desirable and eco-comfortable development strategy evolved through a series of action plans at Panchayat level with people's participation. One of the objectives of PRM was to prepare comprehensive local [panchayat] area action plans for sustainable development with active participation of local people. [For a detailed account of the strategy of PRM and activity linkages, see CESS et al,1991. See also 'Kalliasseri Experiment in Local Level Planning', IRTC 1995. Kalliasseri was one of the 25 Panchayats selected for the project/mapping exercise. The significance of the PRM Programme in Kalliasseri was that data collection and interpretation graduated to the stage of formulation of a comprehensive village plan. Also see Muralikrishna, 2001.
26. With institutionalization, LSGs in Kerala perform a range of functions like:
    - Informer [just providing information to government].
    - Consultant [tendering advise and opinion to government].
    - Agent [operating on behalf of government without any real decision making].
    - Manager [managing institutions and services on behalf of the government].
    - Partner [with government doing the other part].
    - Actor [playing the role of an autonomous agency deciding priorities and resource allocation by itself] [Vijayanand, 2001].
27. According to the statistics compiled by the Institute of Social Sciences, Documentation and Information Centre, among the different states, only Kerala, Madhya Pradesh,

Rajasthan and West Bengal have instituted DPCs in all the districts as on 1 October, 2001. See, PR Update, ISS, Vol. VIII, No. 10,94, October 2001.

28. It is after five years of the enabling provisions made in the PR and Municipality Act that the Appellate Tribunal has been set up. To start with, one Tribunal has been set up at Thiruvananthapuram.

29. Citizens' Charter that came into effect on May 17, 2004 is a statement of services, paid and unpaid, that an LSG offers to the citizens for improving their living conditions. It includes the issuance of permits, licenses and certificates for different purposes. A Charter usually enlists the particulars of services, conditions for obtaining those services and the time limit within which the services must be made available to the citizen.

30. The Examiner Local Funds Audit, who is required by law [Section 215 of Kerala Panchayati Raj Act, 1994] to certify the accounts of LSGs and in fact vested with all powers of civil courts under the Code of Civil Procedure 1908, did not certify the accounts. The state government failed to submit the certified accounts to the legislature and to publish them (Mukhopadhyay, 2005). However, it is a matter of relief that a supplementary Report by CAG of India on the LSGs, under the scheme of technical guidance and support to the Examiner Local Fund Audit for the year ended March 31, 2004 was submitted to the state legislature, which constituted a committee on local funds accounts in June 2003.

REFERENCES

Alsop, R., Krishna, A and Sjoblom, D (2000): "Are Grama Panchayats Inclusive?" Report of a Study Conducted in Rajasthan and MP, Background Paper No. 3, World Bank Unpublished Project Work, *Overview of Rural Decentralization in India*, Vol. 3.

Beier, Christoph (1999): "The Impact of Decentralization on Rural Development", *Agriculture and Rural Development,* Vol. 6, No. 1, pp. 18-21.

Campbell, T., Peterson, G and Brakarz, J (1991): "Decentralization to Local Government in LAC: National Strategies and Local Response in Planning, Spending and Management", Latin America and the Caribbean Technical Departments, Regional Studies Programme, Report No. 5, Washington D.C: World Bank.

CESS et al (1991): *Panchayat Level Resource Mapping: An Approach Paper- A Model for Micro Level Resource Survey with People's Participation.*

Cheema, G.S and Rondinelli, D.A (1983): *Decentralization and Development: Policy Implementation in Developing Countries*, Cheema and Rondinelli (eds), Beverly Hills: Sage.

ESCAP, UN (1991): "Fiscal Decentralization and the Mobilization and use of National Resources for Development: Issues, Experience and Policies in the ESCAP Region", *Development Papers* 11, pp. 1-37.

Giridharan, M.P. (2001): "Involvement of Women in Planning, Implementation and Monitoring of Projects Under People's Campaign for Decentralized Planning", *Evaluation Studies on Decentralized Planning*, State Planning Board, GOK.

Government of India (2000): *Report of the Eleventh Finance Commission*, New Delhi.

Government of Kerala, State Planning Board (1991): *Report of the Steering Committee on Decentralized Planning and Planning for Rural Development*, Eighth Five Year Plan 1990-95.

Government of Kerala (1996a): Ninth Five Year Plan–Decentralization and People's Participation–Mass Campaign, GO. (MS) No. 10/96/plg Dated 30th July 1996, Planning & Economic Affairs (A) Department.

Government of Kerala, State Planning Board (1996b): *People's Campaign for 9th Plan: An Approach Paper.*

Government of Kerala (1996c): Ninth Five Year Plan–People's Campaign for Formulation of Plan–High Level Guidance Council–Constituted, GO (MS) No. 12/96 plg dt. 8th August.

Government of Kerala, State Planning Board (1999): Economic Review 1999.

Government of Kerala, State Planning Board (2001): *Evaluation Studies on Decentralized Planning: Summary of Reports.*

Government of Kerala (2002): Guidelines for the Preparation of Tenth Five Year Plan by Local Governments, GO (MS) No. 20/2002/plg. Dated 6.6.2002 Planning and Economic Affairs (A) Department.

Government of Kerala, State Planning Board (2005): Economic Review 2004.

Harilal, K.N (2005): "New Vistas of Local Governance: A Study of Democratic Decentralization in Kerala", Paper presented in the International Conference on 'A Decade of Decentralization in Kerala: Issues, Options and Lessons', Organised by Institute of Social Sciences, New Delhi, in Thiruvananthapuram, 7-9, October 2005.

Institute of Social Sciences (2001): Panchayati Raj Update, Vol. 11, No. 10, 94.

Issac, Thomas T.M and Harilal, K.N (1997): "Planning for Empowerment: People's Campaign for Decentralized Planning in Kerala", *Economic and Political Weekly*, Vol. XXXII, No. 1 & 2, pp. 53-58.

Jha, S (2000): "Fiscal Decentralization in India: Strengths, Limitations and Prospects for PRIs", Background Paper No. 2, World Bank Unpublished Project Work, *Overview of Rural Decentralization in India*, Vol. 3.

Kaushik, P.D (2005): "Panchayati Raj Movement in India: Retrospective and Present Status", in Bibek Debroy and P.D. Kaushik (eds) *Energizing Rural Development Through Panchayats*, Academic Foundation and RGICS.

Kerala State Planning Board (1997): The People's Campaign for Ninth Plan in Kerala: An Introduction.

KRPLLD, Center for Development Studies (1995): Kalliasseri Experiment in Local Level Planning, Seminar on Local Level Planning-Lcarning from Kalliasseri, 2-3, October.

KRPLLD, Center for Development Studies (2003): Annual Report 6 (2001-02): Panchayat Level Development Planning: Problems and Prospects of People's Participation in a Party-dominated Democracy, pp. 27-44.

Maetz, Materne and Quieti, Maria G (1987): "Training for Decentralized Planning: Lessons from Experience", Vol. 1, *Development Policy Studies and Training Service*, Policy Analysis Division, FAO.

Mathew, Anand (2005): "Resource Mobilization and Intergovernmental Transfers", Paper presented in the International Conference on 'A Decade of Decentralization in Kerala: Issues, Options and Lessons', Organised by Institute of Social Sciences, New Delhi, in Thiruvananthapuram, 7-9, October 2005.

Mathew, George (2001): "Panchayat Elections: Dismal Record", *Economic and Political Weekly*, 20 January.

Mathew, George (2003): "Panchayati Raj Institutions and Human Rights in India", *Economic and Political Weekly*, Vol. XXXVIII, No. 2, pp. 155-162.

McLean, Keith (2000): "Political Decentralization: International Experience', paper presented in the IMG Seminar, Thiruvananthapuram, Kerala.

Minis, Henry and Dennis, Rondinelli (1989): "Promoting Economic Development and Employment Generation Through Decentralization in Senegal", USAID Working Paper, mimeo, p. 122.

Mohanan, S (2001): "Resource Mobilization in People's Planning: A Case Study", *Evaluation Studies on Decentralized Planning*, State Planning Board, GOK.

Mukhopadyay, Amitabh (2005): Accountability and Communicative Action: An Alternative Imaginary, *Economic and Political Weekly*, Vol. XL. No. 52, pp. 5463-69.

Muralikrishna, M (2001): "Evaluation of Panchayat Level Resource Mapping for Panchayat Level Planning", *Evaluation Studies on Decentralized Planning*, State Planning Board, GOK.

Narayana, D (2005): "Local Governance Without Capacity Building: Ten Years of Panchayati Raj", *Economic and Political Weekly,* June 25, pp. 2822-32.

Oommen, M.A (2004): Preface to "*Fiscal Decentralization in Kerala*", in Geeta Sethi (ed) *Fiscal Decentralization to Rural Governments in India*, The World Bank, Oxford University Press.

Parker, Andrew N (1995): "Decentralization: The Way Forward for Rural Development?", *Policy Research Working Paper* 1475, Agriculture and Natural Resources Department, Sector Policy and Water Resources Division, The World Bank.

Pillai, P.P (2005): "Capacity Building for Decentralized Democratic Governance: Some Lessons from Kerala Experience", Paper Presented in the International Conference on 'A Decade of Decentralization in Kerala: Issues, Options and Lessons', Organized by Institute of Social Sciences, New Delhi, in Thiruvananthapuram, 7-9, October 2005.

Prakash, B.A (2005): "Plan Performance of LSGs: A Study in 36 Grama Panchayats in Kerala", Paper presented in the International Conference on 'A Decade of Decentralization in Kerala: Issues, Options and Lessons', Organised by Institute of Social Sciences, New Delhi, in Thiruvananthapuram, 7-9, October 2005.

Prudhomme, Remy (1994): "On the Dangers of Decentralization", *Policy Research Working Paper* 1252, Transportation, Water and Urban Development, The World Bank.

Putnam, R (1993): *Making Democracy Work*, Princeton: Princeton University Press.

Rondinelli, D (1981): "Government Decentralization in Comparative Perspective: Theory and Practice in Developing Countries", *International Review of Administrative Science*, 47.

Rondinelli, D.A (1983): "Decentralization of Development Administration in East Africa", In Cheema and Rondinelli 1983, pp. 77-125.

Rondinelli, Dennis and Nellis, John (1986): "Assessing Decentralization Policies in Developing Countries: The Case for Cautious Optimism", *Development Policy Review,* 4:3-23.

Sahay, Gaurang R (2005): "Decentralization, Panchayati Raj and the Rights of Marginalized Groups in India: A Case Study for Civil Society", Paper Presented in the NIRD Foundation Day Seminar 9-10, November 2005, National Institute of Rural Development, Hyderabad.

Sankaran, P.N (2001): "Towards an Artisan Plan", *Evaluation Studies on Decentralized Planning*, State Planning Board, Government of Kerala.

Sankaran, P.N. (2005): "Towards Transparency and Corruption-free Governance: An Overview of Decentralization in Kerala", in A. Ranga Reddy [ed] *Dimensions of Crime and Corruption in India,* Serials Publications, pp. 160-185.

Shah, Anwar and Thompson, Theresa (2004): "Implementing Decentralised Local Governance: A Treacherous Road with Patholes, Detours and Road Closures", *Panchayat Raj Update*, Vol. X1, No.1,121, January, pp. 6-7.

Stockmayer, Albrecht (1999): "Decentralization: Global fad or Recipe for Sustainable Local Development", *Agriculture and Rural Development*, Vol. 6, No.1, pp. 3-21.

Sudhakaran, M.N (2004): "Plan Integration in the Context of Multi level Planning–A Study based on Thrissur District", *Discussion Paper* No.102, Kerala Research Programme on Local Level Development, Center for Development Studies, Kerala.

The World Bank (2002): "Decentralization in Kerala", Chapter III, *Rural Decentralization in India.*

The World Bank (2004): Fiscal Decentralization to Rural Governments in India, [ed] Geeta Sethi, Oxford University Press.

Thomas, Baby (2001): "The Impact of Participatory Decentralized Planning on the Nature of Projects and Own Resource Generation of Panchayats in Kerala", *Evaluation Studies on Decentralized Planning*, State Planning Board, GOK.

UNDP (2002): Human Development Report 2002, *Deepening Democracy in a Fragmenting World*, Oxford University Press.

Vijayanand, S.M (2001): "Issues Related to Administrative Decentralization and Administering of Decentralization: Lessons from the Kerala Experience", paper presented at the Workshop on Decentralization, ISEC Bangalore, 31st May and 1st June.

Vijayanand, S.M (2002a): "Status of Panchayati Raj in Kerala", Paper presented in the IMG Seminar, Thiruvananthapuram, Kerala.

Vijayanand, S.M (2002b): "Fiscal Decentralization in Kerala: Points for Discussion", Paper presented in the IMG Seminar, Thiruvananthapuram, Kerala.

Wadia, F.K (1996): "Panchayaths in India–Approaches and Ground Realities in a Historical Perspective", *Journal of Indian School of Political. Economy*, Vol. VIII, No. 1, pp. 64-97.

White, Roland and Smoke, Paul (2005): "East Asia Decentralizes", in East Asia *Decentralizes: Making Local Government Work*, The World Bank, Washington D.C., pp. 1-23.

Part - II

DECENTRALISATION AND THE DYNAMICS OF INSTITUTIONS

2

INSTITUTIONAL REFORMS FOR PANCHAYATI RAJ: TOWARDS LINKING COOPERATIVES WITH LOCAL GOVERNMENTS

THE PROBLEM

There is a growing interest in decentralized planning and implementation among developing countries. Transfer of planning and administrative authority from the central government to sub-national levels has already been undertaken in diversified ways and with uneven degrees of success. Decentralized planning is understood as the result of an overall institutional effort undertaken by a country to decentralize decision-making according to a process defined throughout the country (Maetz and Quieti, 1987). Accordingly, this concept differs from 'local-level planning', which appears more as a result of local, often isolated initiatives, without special institutional setup or links with the national level. Despite considerable theoretical and empirical exercise on decetralisation with reference to the experiences of countries in Asia, Latin America, Near East and North Africa, South of Sahara etc., the institutional dimensions of the process have often been given insufficient attention, due to the difficulty associated with characterizing the role that Non-Governmental Organisations (NGOs) and other community-based organizations might play in decentralization. The result is that no framework has been identified that assesses all the dimensions of decentralisation and their interlinkages (Parker, 1995).

Concern with the institutional dimensions of development is a relatively new phenomenon. This is clearly reflected in the earlier decentralization initiatives in Africa and Asia that focused almost entirely on the administrative institutions of government. In the decentralization experiments of the late 1980s and early 1990s, the scope of NGOs, co-operatives and other grassroots organizations in the process was still perceived as peripheral (Parker, 1995). Hence the next task of theoreticians and development practitioners of decentralization lies in the identification of linkages of local governments on the one hand with third sector[1] institutions (like NGOs, co-operatives, voluntary agencies) and self-help groups on the other. Third sector operates between the public and the private sectors. It can be designated variously as the voluntary sector, the membership sector, the self-help sector, the participatory sector or the collective action sector. In the decision making process, it depends more on voluntaristic accommodation and persuasion with reference to both group and individual interests (Hunter, 1969 and Uphoff, 1993).

Among the 'third sector' institutions who have been increasingly associated with rural development and decentralized planning, the co-operatives have emerged as the most suitable agencies as they are more popular, closer to the rural population, people-centred and user-controlled, compared to NGOs and voluntary groups. A co-operative is a legally recognized form of association of persons, owned by its members, whose main principles include voluntary membership to achieve a common end and equal voting rights for each member. Despite past problems with co-operatives, they remain , in many countries, the preferred form of organization with which rural people are familiar (Hussi et al, 1993). The International Co-operative Alliance in the Congress held in September 1995 has accepted 'Concern for Community' as the seventh principle in the statement on the New Co-operative Identity. The co-operatives have assumed an added focus in the context of the ongoing dismantling of the regulated marketing and pricing systems, and the decentralization of the planning process. Participatory development at the community level is gaining importance and co-operatives are being rediscovered (Levi and Litwin, 1986). Given the potential of rural co-operatives in the promotion of socio-economic development, their linkage with local governments like Panchayats in India may add a new dimension to the content and practice of decentralized planning. It may be remembered that despite the considerably long history of Panchayati Raj Institutions (PRIs) and decentralized planning, the record of the Indian experiment in the field has been far from satisfactory.

THE CONTEXT

The preparation of plans for economic development and social justice is an important function entrusted to Panchayats[2] under Article 243 G of the Constitution of India which was introduced in Part IX by the 73rd Amendment Act, 1992. Elections to Panchayats at the village, block and district levels have become mandatory; most of the States have made enactments for the purpose[3] and PRIs have come into existence throughout the country (see Annexure 1 for the Kerala scenario). In contrast to PRIs that existed in the country from time to time, the present system is stated to have forces of strength, viz, constitutional status and backing, ensured structure, periodicity of elections, representation for weaker sections and possibilities of reservation for backward classes. Opportunities for undertaking integrated local development, mobilization of local resources, self-help and community cooperation, involvement and participation of people etc., also exist in the new system (Ramachandran, 1993).

Among the advantages of PRIs and decentralized planning, the concept of people's or popular participation remain rather nebulous and difficult to operationalise. In view of economic reforms with diminishing State and expanding markets, what is the appropriate institutional mechanism that can facilitate people's involvement in local level development? We presume that it consists mainly of the Third Sector option namely NGOs, grassroots organizations, voluntary agencies, co-operatives, etc; of these, the co-operatives (especially the credit societies that operate at the grassroots level) constitute a democratically organized potential agency, that can promote people's involvement in plan formulation, implementation and monitoring. In regions like Kerala, the grassroots co-operatives like Primary Agricultural Credit Societies (PACS) cover all villages, operate closer to the village population and possess the highest level of deposits, membership, share capital etc., per society, compared to the national average (see Annexure 2). The above concerns/considerations have defined our agenda of linking PACS with Panchayats in the Kerala scenario.

PANCHAYATI RAJ, DECENTRALISATION AND THE INSTITUTIONAL ISSUES

PRIs in India refer to a statutory, multi-tier administrative structure entrusted with developmental duties and responsibilities by the State legislature. Pandit Nehru inaugurated the Panchayati Raj at Nagaur in Rajasthan on October 2, 1959, following the acceptance by the National Development Council of the Balwantrai Mehta Committee recommendation that public participation in community works should be organized through statutory representative bodies. The advocates of the system wanted PRIs to be vibrant institutions of self-government. The Panchayati Raj and decentralized planning process have been viewed as part of a 'development vision' which conceives organizations such as NGOs and co-operatives at the grassroots level as instruments of people's participation[4] capable of enhancing democracy, social justice, self reliance, sustainability and the elimination of exploitation in development programmes.

In the evolution of PRIs in India, the role of institutional setting conductive to their development has received scanty attention from the policy side. It is true that, until recently, mainstream economic theory has by and large ignored institutional issues, often stating central propositions with a false air of institutional neutrality (Bardhan, 1989). However, in the last few decades, institutional weaknesses have been referred to as critical constraints in many programmes for economic development.[5] The World Bank notes that the development community has become increasingly aware of the need for institutional strengthening[6] if development is to be sustained (Gray et al, 1990). Hence it is obvious that the success of grassroots democracy and development envisaged under the Panchayati Raj depends much on the strength and linkages of local institutions, people's participation and empowerment.

The local level institutions supportive of decentralized planning are the voluntary agencies, community groups, NGOs[7], co-operatives, self help groups etc. Unfortunately, even after the landmark 73rd Amendment to the Constitution of India providing statutory basis to the PRIs, no serious effort has been made towards strengthening the institutional input. Consider for example the condition of co-operatives who can serve as the most potent grassroots organization in rural

India. Most of them are either under the grip of powerful caste lobbies or under government administration/liquidation with little scope for member participation and education or autonomy in operation.[8] Deficiency in institutional preparedness has by and large plagued the pace of Panchayati Raj in several parts of the country at different stages. Administrators and academicians have emphasized this vital institutional environment for the survival and success of the Panchayati Raj in India (Krishna, 1994). Promotion of the right type and mix of institutions that interact with the PRIs with necessary incentives and accreditations, should have received top priority in the agenda of democratic decentralisation and Panchayati Raj.

A prerequisite for increased involvement of local governments in local development is a meaningful form of decentralisation, which would give the local level more leeway in investment decisions. Decentralization in the economic sense implies the transfer of responsibility for planning, management, resource raising and allocation of resources in the public sector from the central to lower authorities or units (United Nations, 1962; Rondinelli, 1983).[9] Besides the Rondinelli terminology, a further analytical refinement can be discerned in the literature between political, fiscal and institutional dimensions of decentralization.[10] Conyers (1990) has listed five arguments in favour of decentralisation relating to planning. Decentralization would:

- increase popular participation in planning and development
- make plans more relevant to local needs
- facilitate co-ordinated or integrated planning
- increase the speed and flexibility of decision making, and
- generate additional resources and encourage more efficient use of existing resources.

The objectives of decentralization as put forward by some countries are:

- promotion of a more balanced development in the country
- design of more realistic projects and programmes which take into account local potentialities and constraints
- more effective co-ordination of development activities at various spatial levels through disaggregation of planning function
- Strengthening of local political institutions and increase of people's participation in development; and
- mobilisation of local resources (Maetz and Quieti, 1987).

Eventhough decentralized planning for rural areas has attracted considerable political and academic attention, the record is bleak. Central governments often consider decentralized planning as a conspicuous effort to erode their power base and point at inefficiencies and incompetence at local levels. Local governments complain about the lack of appreciation of local situations by central government. It has often resulted in somewhat lethargic attitudes on the apart of local governments and decentralized planning being considered as an obligatory ritual to be performed, rather than as means to promote and guide development in a rural area (Herlaar and Olthof 1994).

While decentralized institutions may be more susceptible to local participation and control, by the same token they may be captive to the local structure of power (Arkadie, 1989). It is also maintained that decentralization may result in allocative and productivity inefficiency apart from making redistribution and macro-economic stabilization policies more difficult (Prudhomme, 1994). The drawbacks of centralized power–insensitivity to local needs and ignorance of local capacities–have their counterparts in the skewed access that can arise from local systems of power and prejudice (Arkadie, 1989). With more power and resources, and less accountability, PRIs may become a bed of harassment of the SCs, STs, and women who have no protection. Similarly, funneling of resources and powers from the centre on a vast scale to PRIs which are not prepared and experienced to use them in a proper and competent way may create massive problems of misuse, wastage and financial mismanagement. In a way, corruption along with power will be decentralized if institutional degeneration is not reversed and supportive institutions not promoted.

The current and foreseeable trend towards a diminished role of the State in most economic activities, including rural development offers new opportunities for local organizations and institutions. Hence the problem of Panchayati Raj is not mainly inadequate decentralization of economic planning, as it is often made out to be, but of redefining State-PRI relations, reforming the rural socio-political scenario, providing the right type of incentives for institution building and better PRI-collective action sector interface.

LOCAL GOVERNMENTS AND GRASSROOTS INSTITUTIONS: THE EMERGING FOCUS

The experience of less developed countries in rural economic development over the last half a century had demonstrated that neither direct state intervention nor indirect state influence through macro policies is a sufficient condition for success. The failures in such approaches and efforts often lie in communication and implementation lapses between households, markets and the State. Such failures can be attributed to the lack of vibrant functional and healthy local organizations and institutions and can be overcome by their presence (Korten, 1980; ILO, 1990 and Nugent, 1993).

Many developing communities are therefore facing the reality that big government (and big business) cannot resolve satisfactorily their economic problems of growth and equity. In the context of contemporary global economic restructuring, the nation-state is losing much of its capacity to promote the well-being of its cititzens (Cooke, 1990, 1995). Accordingly, as economic development has become increasingly a localized phenomenon, significant shifts are occurring in the locus of responsibility for development planning from national to sub regional or local levels (Ettbinger, 1994). As Sengenberger observes, the "local level"[11] is exerting "an increasingly strong attraction on analysts and policy makers today" across a wide political spectrum of opinion (Sengenberger, 1993).

With the decline of the nation-state, the region or locality therefore emerges as the relevant economic space "in and for which co-ordinated efforts to promote wealth-creating economic activities can be meaningfully undertaken" (Cooke, 1995). The diverse nature of local government and planning responses has included variously the support of small firms, co-operatives, self-help networks or support agencies. Local governments can therefore enter into new partnership arrangements with private or community-based organizations like co-operatives and assume an initiating rather than passive role. This will have implications on both institutional arrangements and on the roles, functions and skills of local development planners.

The view that local governments could play a more positive role in development process predominates today; this has lead to pressures for more decentralization. Potentially strong points of local governments emerge from two issues [see also Mawhood, 1983; Olowu, 1988]. First they are closest to the people and thus have a detailed knowledge of their needs, wishes and demands. They are also in a strong position to communicate and explain, and to incorporate local views and ideas. This not only contributes to responsiveness and accountable governance on the part of local government, but also to a sense of 'ownership' if decisions are felt to be taken by [i] decongesting them, [ii] providing a check on their performance, [iii] augmenting the effective tax base, and [iv] cushioning antagonisms between central government and local movements [Herlaar and Olthof 1994].

Eventhough the concept of 'Local Government' is both fuzzy and varies by country, certain institutional characteristics may be noted. They involve combinations of several features, viz., democratic elections, specific functions based on law, power to tax, power to decide on and operate own budgets, power to hire [and fire] own staff and power to issue byelaws [Nathan et al, 1994]. Mawhood has characterized a decentralized local authority as "having its own budget, a separate legal existence and the authority to allocate substantial resources on a range of different functions, the decisions being made by representatives of the local people" [Mawhood, 1993].

Due to their proximity to the population, local governments are often deemed to be the ideal vehicle for democratizing local institutions and opening them to popular participation from below [Schonwalder, 1992]. Hence there is growing evidence that working with, and through collective action organizations [like co-operatives] at local levels can produce improvements in productivity, and well-being more far-reaching than the stereotype that local efforts benefit only a few communities or a few families [Uphoff, 1988]. As the state is increasingly forced to reduce its funding for rural public and semi-public goods, success in rural development will require an increasingly good match between local organizations and institutions and the incentives for the creation and maintenance of these goods [Nugent, 1993]. Therefore the rationale of linking primary co-operatives [one of the most popular grassroots organization] with panchayats [being a unit of local government] has to be viewed against the above theoretical foundations of current thinking.

OBJECTIVES OF THE STUDY

The broad objective of the study is to identify between local governments [Panchayats] and cooperations (PACS) the area of co-ordination required to enhance people's participation in development planning. In view of local needs, priorities and resource constraints of PRIs on the one hand and business interests, democratic identity and user control of PACS on the other, how the latter can contribute to the formulation and implementation of plans by PRIs is a related objective of the study. It is also the objective to sort out the changes in the organizational frame of PACS that might be conducive for them to function as sustainable institutions supportive of the Panchayati Raj. With the above objectives in view we have undertaken an expert opinion survey in Kerala during April–May 1996. The present paper mainly reports on the findings of the opinion survey.

METHODOLOGY

(a) Conceptual and Theoretical Background

Common membership of the village population in the PACS and Grama Sabha[12] has been conceptualized as the framework for operationalising the proposed linkage between the PACS and PRIs. The larger the operations and membership in the PACS the larger is its potential as a mechanism for promoting peoples participation in Grama Sabha and Panchayat Level planning. Though PRIs, being local governments, come under the definition of the state sector with its mode of operation 'top-down', decentralization of planning has reversed the process to 'bottom-up' which is typical of institutions like PACS in the third sector. The analogous mode of operation in PRIs and PACS (despite differences in institutional categories) through the process of people's/member participation perhaps lays one of the theoretical foundations for the proposed linkage between the two.

In terms of functions performed, the village co-operative promotes thrift, and mutual help and advances funds for agricultural and rural development ventures/ activities. The Panchayat, based on micro plans initiated at the Grama Sabha level, prepares comprehensive development plans for the area in annual and long–term perspectives. Though the operations of the PACS through the instrument of the 'credit cycle'–collection/mobilization of deposits, lending and recovery-promote directly productive economic activity of the local area, their lending programmes are by and large credit–linked and are undertaken not as part of Panchayat plans. Development banking consists of enlarging the dimensions of the credit cycle besides fine-tuning and eliminating interruptions in the mechanism. One such dimension is the linkage of its functions with comprehensive Panchayat development plans.

Decentralised and local level planning, of course, demands much more from PACS than what is normally conceived and summarized in their operations. Many a time lending operations of PACS, as already said, may not be for the type of activities as prioritized in panchayat plans. Similarly, programmes for deposit mobilization may not be linked with financing of panchayat plans. Inadequate development of economic infrastructure is a major constraint that may hold back development in the directly productive sectors (for which the PACS provide loans and advances). Therefore, infrastructure development should be high on the agenda of the panchayat plan. Moreover, the panchayat is charged with a number of welfare functions also as implementing agency for central sector schemes. Hence, the operations of the PACS should have meaningful and scientific linkage with panchayat plans as the two institutions are complementary and mutually reinforcing; the PACS stand out as a promising vehicle for the transmission of people's participation in planning.

(b) Method Used

The methodology adopted for the study on linking the PACS with PRIs in plan formulation and implementation has been the feedback and results of a Delphi Study [expert opinion survey]. It is a technique which involves a process in learning the opinions of people who are specially knowledgeable and has been defined as a method for structuring a group communication process so that the process is effective in allowing a group of individuals, as a whole, to deal with a complex problem [Linstone and Turoff, 1975]. Face to face contact between the participants is not required; a series of mailed questionnaires is usually used, each one sent out after the results of the previous one have been analyzed. The elements that are usually included are an opportunity for individuals to contribute ideas or information or assessment of the group judgement, classification of reasons for difference, a chance for individuals to revise their views, and some degree of anonymity for the individual responses. Votes may be cast and results fed back repeatedly until stability or consensus is reached.

(c) The Procedure

For the present study, we started the opinion survey with a structured questionnaire that was prepared after preliminary informal interaction with a handful of experts in co-operation and local government. The panel of respondents comprised experts in the field of local government and administration [31%], co–operation and finance [36%], training, teaching and research [19%], and local planning, monitoring and evaluation [14%]. In the first round, 64 experts inclusive of non–officials responded to the mailed questionnaire. Experts were asked to state their opinion on various issues by using a five point scale rating, viz., [1] Strongly agree, [2] Agree, [3] Cannot answer, [4] Disagree, and [5] Strongly disagree. We limited the round of responses to two, since consensus was reached on important issues at that stage. In the second round of questionnaire, we could collect the responses of 51 experts. Some of the statements which failed to elicit consensus were redrafted and sent to the experts. A resume of the statements on which consensus had been reached were also placed before the experts along with the majority opinion and or minority views. The experts' response in the first and second round is analyzed and expressed in terms of three quartiles. The average opinion of the panel is reflected in the second quartile that takes into account each opinion. The semi–interquartile range is used to show the spread of opinion. In this endeavour, consensus is said to have been arrived where the semi–interquartile range is less than or equal to unity.

RESULTS AND DISCUSSION

Acknowledging the potential that underly the principles of common membership, complementarity and mutual reinforceability, as outlined in methodology, the experts were of the opinion that a meaningful linkage can be set up without loss of autonomy in the PACS and PRIs. Irrespective of their official or non–official status, the panel strongly agreed on the need for strengthening people's participation through institutions like the PACS for the promotion of the Panchayat Raj. Experts strongly shared the view that there has been very little initiative towards linking grassroots co-operatives with local governments/PRIs, retaining their distinct institutional identities/autonomy. Similarly, experts endorsed the opinion [elaborated in ILO, 1992] that the co-operatives, who possess an organizational structure suited to serve the weaker sections of the community, have greater significance in view of the current phase of economic reforms- privatization, liberalization and globalization. However, on the academic and widely shared non-official view [of organizations such as Co-operative Development Foundation] that co-operatives are highly bureaucratized/politicized with a paralyzing effect on genuine people-centered and user-controlled co-operative development, there was no consensus. The panel, however, agreed on the need for co-operatives to function also as business concerns, especially in view of the emergence of the non-farm sector in the state's economy.

The pace of the Panchayati Raj has suffered setbacks in the past, mainly on account of the mismatch between financial decentralization [due to the lukewarm official attitude to devolution of funds] and functional decentralization. In a recent paper, Rajaraman has argued that there has to be some revenue additionality to cover incremental extablishment costs in the Panchayati Raj [Rajaraman et al, 1996]. In response to the question on potential sources of local finance, there was agreement on the existence of surplus funds in the co-operative credit system of Kerala. Panelists agreed that the proposed linkage between PACS and PRIs would facilitate better local resource mobilization. However, the resource potential argument based on the surplus funds or excess liquidity syndrome[13] of PACS was not admitted as a sufficient ground for reduced dependence of PRIs on government grants, loans and tax devolution. Opinion was sharply divided on the prospects of co-operatives floating independent subsidiaries[14] and their involvement in commercial and infrastructure development activities as part of the Panchayat Plan. PRIs already have the option to keep their funds with government treasury, commercial and co-operatives banks. The panelists favoured PACS in order to strengthen the ties between the two. There was also consensus on the proposal to deploy larger volume of funds by PACS for approved schemes of the Panchayat Plan and linking of credit plan of the PACS with the development plan of the Panchayats.

Agreement on the above broad considerations provided the schematic framework for the operationalisation of the proposed Panchayat-PACS linkage.

The mechanism endorsed by the expert panel is a 'Joint Planning and Coordination Committee' (JPCC) of the Panchayat, PACS, Development Departments and Volountary Agencies (of the locality) in which the Chairman shall be the President of the Panchayat, Vice-Chairman the President of the PACS and the Secretary of the Panchayat and PACS to be the Convener and Joint Convener respectively (see Figure 1.

Comprehensive Panchayat Plan for the development of human, land and environmental resources, promotion of traditional arts and crafts, cottage and village industries, housing and infrastructure, water supply etc., may be formulated and implemented by the JPCC. The JPCC shall also have powers for the conduct of socio-economic survey, trade fairs, sports festivals and for the promotion of arts, crafts and culture. Moreover, the JPCC shall have authority to undertake special resource mobilization campaigns and recovery drives for co-operative advances.

Given the multi-agency system of banking and 'third sector' institutions at the local level, it was agreed that the JPCC shall have special invitees, as and when required, consisting of the representatives of Commercial, Agriculture Development and Regional Rural banks, Community Service Institutions etc. The suggestion that the bond between Panchayat and PACS can be further strengthened by admitting the former as an institutional member in the latter (without voting rights) by appropriate amendments to laws was agreed to by the panelists. Similarly, in view of their prominence among grassroots organizations, there was agreement to provide permanent membership of the PACS in the JACS. However the panelists brushed side a suggesting for amendments to laws and rules to provide for mutual participation (without voting rights) of two representatives each (the secretary and a director/member) in the executive/standing committee meetings of the Panchayat and PACS. The suggestion was turned down on grounds of encroachment on the autonomy of the respective institutions. Experts agreed on the need for periodic training, and workshops in decentralized planning, local administration, financial management, community organization etc. for officials and non-officials in the PACS and Panchayats.

CONCLUSIONS AND POLICY IMPLICATIONS

The 2,57,000 rural bodies that have been mandated under the 73rd Amendment make this perhaps the largest experiment in decentralization of governance attempted anywhere. Whether it has come about because decentralization is currently fashionable internationally (Tanzi, 1995; Bird, 1994) or because of internal compulsions, a great deal hinges on the success of it. We maintain in the present paper that its success is very much dependent on the operationalisation of people's participation in planning through linkages with grassroots organisations which are closer to the village population like PACS.

The Delphi method used in the present study is of an intuitive nature based on an expert opinion survey through a process of iteration. The technique has helped to reach an agreement or consensus on a set of statements covering various aspects of the linkage problem. The process of obtaining responses with anonymity among

Figure 1
Schematic Framework for Panchayat Level Planning / Development

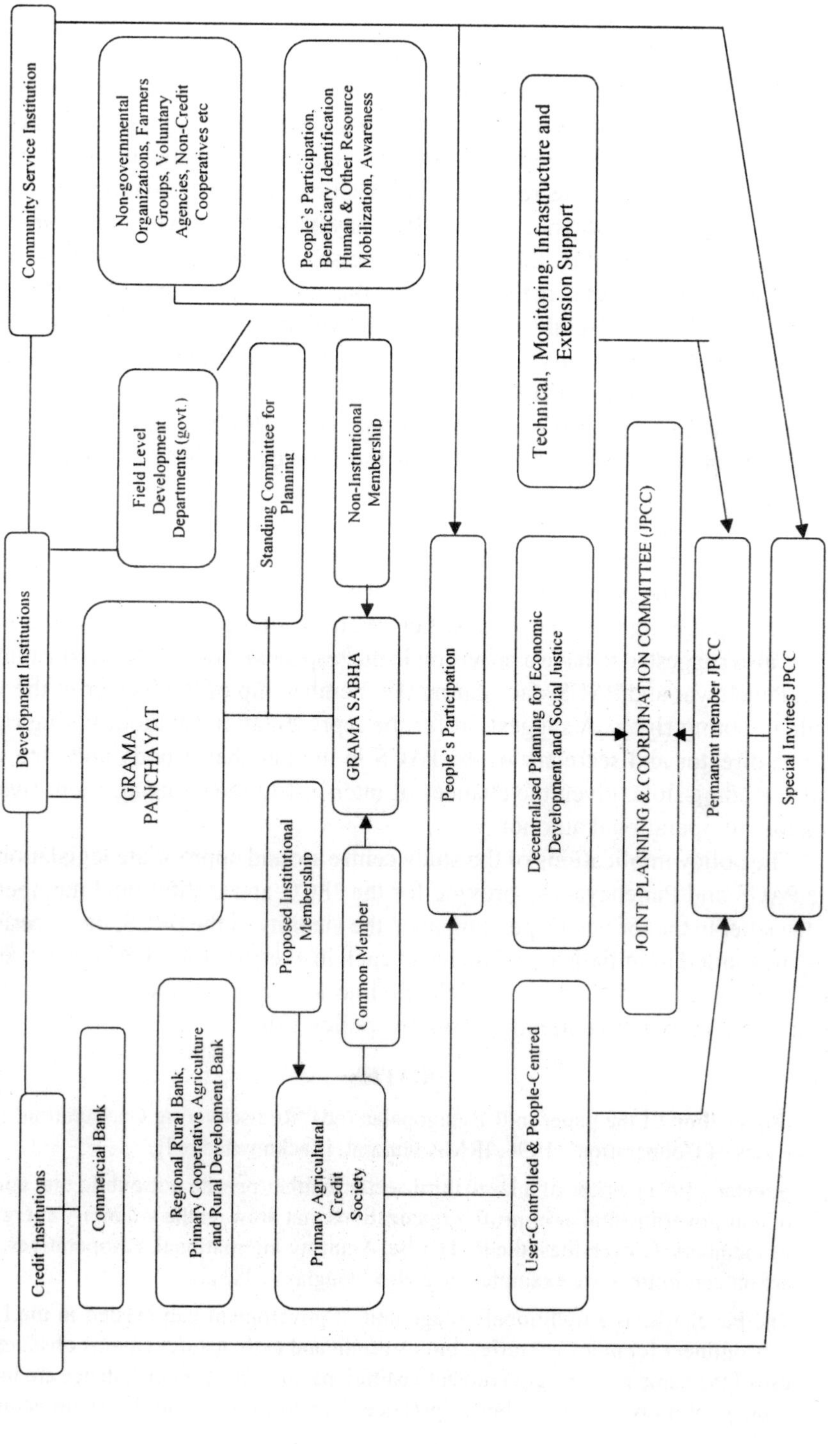

the panelists avoids the possibility of identifying a specific opinion with a particular person. This facilitates a respondent to deviate while responding to the statements of others without losing the face. A number of conclusions of the study and their policy implications are outlined in this part.

PACSs have been accepted as one of the useful agencies to translate the principle of people's participation in Panchayat level planning, given the primacy of PACS as genuine development-oriented autonomous associations of people. Their ability to accelerate resource mobilisation for local development and to function in broad consonance with the Panchayat Plan objectives and perspectives has been widely shared by the panelists. The strategy for deposit mobilisation of PACS has to be formulated in the light of the financial requirements of the Panchayat Development Plan (preferably in consultation with the JPCC); during the campaign for deposit mobilization, its linkage with the Panchayat Plan has to be highlighted with a view to make it a people's programme.

The mechanics of the proposed Panchayat-PACS linkage towards the development perspective, as endorsed by the experts, consists of a Joint Planning and Coordination Committee (JPCC) with representatives drawn from credit, development and community service institutions. As a permanent member of the JPCC, the PACS can reflect on savings potential, activity and technology preferences of the people, gaps in local infrastructure etc., which provide useful insights to planning with people's participation. To strengthen the ties in the JPCC, it was also suggested to have a provision in the respective laws of the two institutions viz., Panchayat and PACS, facilitating the membership of the former in the latter without voting rights. A suggestion for the representation (without voting rights) of one director and secretary of the PACS in the Panchayat body and vice versa was not admissible, given the chances of inter-institutional management rivalries and loss of operational autonomy.

The policy implications of the study centre around appropriate legislation's in the PACS and Panchayats to provide for the JPCC and institutional membership of the latter in the former. Depending upon the viability of the PACS, the experiment may be limited to initially one Panchayat each in every district. Periodical training and conduct of workshops for official and non-official members of the two also requires corresponding attention from the policy side.

NOTES

[Publication of the paper in R.Rajagopalan (ed) 'Rediscovering Cooperation, Vol. 1 Bases of Cooperation', 1996, IRMA Gujarat, is acknowledged].

1. Streetan (1991) writes of a large third sector, neither private nor public that consists of non-governmental, non-profit organizations that draw on the voluntary energies of its members. Universities, the Red Cross, Amnesty International, co-operatives, local action committees are examples. See also Douglas (1983).
2. The Panchayat as a traditional village unit of government had existed in the Indian subcontinent for many centuries, but its origin and early nature remains obscure. The use of the name for local government institutions in the post-independence era implies some continuity with past, but in practice their form, role and the socio-economic

and political context of operation have all significantly altered as the state of India has come into being (Webster, 1990). In Kerala, the early history of PRI may be traced to the Administrative Reforms Commission 1957, the Kerala Panchayat Act 1960 and the Kerala District Administration Act 1979. For a historical perspective of Panchayats in India, see Wadia (1996).

3. For a critical review of State Panchayat Act, see Voluntary Action Network India (VANI) (1995).
4. See for an account of the developmental role of voluntary agencies, Brett (1993).
5. In a presentation entitled "Collective Choice Perspective on Indian Economy" Mancur Olson said that India is not even close to releasing its potential of development due to lack of conducive institutional factors. See "Redistribution as Means of Economic Reform Mooted", *Financial Express,* January 13, 1996, p. 5.
6. An excellent analysis of the role of institutions may be seen in Israel (1987).
7. For a discussion of the role of NGOs, see Dubey (1995).
8. For roots of the co-operative crisis, see Taimni (1993).
9. Rondinelli has distinguished between four different categories of decentralization deconcentration, delegation, devolution and privatization. For details see Rondinelli (1981).
10. An outline of decentralization for rural development is given by Parker (1995).
11. There is little consensus about what constitutes "the local level". Analytically, there are three local levels, viz., the locality level, community level and the group level. See Uphoff (1986).
12. Grama Sabha is an assembly of voters in a ward or division of a Panchayat that has to meet at least twice ą year to discuss the development issues in the area. Its functions include preparation, promotion and implementation of development schemes, mobilisation of money, materials and manpower for local use, assistance for the identification of beneficiaries of welfare programmes, consideration of development and administration reports, annual statement of accounts etc., pertaining to the ward.
13. Excess liquidity in the co-operative credit system in Kerala is a live issue. See Sankaran (1994).
14. One such suggestion has been made in the context of Kerala. See State Planning Board (1996). Also see Planning Commission GOI, (1991).

REFERENCES

Arkadie, Brain Van (1989): "The Role of Institutions in Development", *Proceedings of the World Bank Annual Conference on Development Economics,* The World Bank, 1990, pp. 153-179.

Bardhan, Pranab (1989): "Comment on the Role of Institutions in Development by Van Arkadie", *Proceedings of the World Bank Annual Conference on Development Economics,* The World Bank, 1990, pp. 181-185.

Bird, Richard (1994): "Financing Local Services: Patterns, Problems and Possibilities", Paper prepared for *Global Report on Human Settlements*, April.

Brett, E.A (1993): "Voluntary Agencies as Development Organizations: Theorizing the Problem of Efficiency and Accountability", *Development and Change*, 24 [2] pp. 269–303.

Conyers, D (1990): "Decentralization and Development Planning: A Comparative Perspective", In P. de Valk and K.H. Wekwete [eds.] *Decentralizing for Participatory Planning,* Aldershot, Avebury.

Cooke, P (1990): *Back to the Future*, Unwin Hyman, London.

Cooke, P (1995): *Reinventing the Region: Firms, Clusters and Networks,* Friedrich Ebert Stiftung, Bonn.

Douglas, James (1983): *Why Charity? The Case for a Third Sector,* Sage Publications, Beverly Hills.

Dubey, Ajay (1995): "Strengthening Panchayati Raj Institutions: Strategies for NGOs", *Working Paper* 87, Institute of Rural Management, Anand, pp. 1-10.

Etttlinger, N (1994): "The Localisation of Development in Comparative Perspective," *Economic Geography,* 70(2): pp. 1444-666.

GOI, Planning Commission (1991): *Report of the Committee on Model Cooperatives Act,* Government of India, New Delhi, pp. 1-57.

Gray, Chery W., Lynn S. Khadiagala and Richard J. Moore (1990): "Institutional Development Work in the Bank-A Review of 84 Bank Projects", Policy, Research and External Affairs *Working Paper*, Development Economics, WPS 437, The World Bank, Washington D.C, pp. 1-27.

Herlarr, Martin and Wim Olthof (1994): "Rural Local Government and Development Planning, with Experiences from Buhera District, Zimbabwe", *Working Paper* Series No. 181, Institute of Social Studies, The Netherlands, pp. 1-42.

Hunter, Guy (1969): "*Modernising Peasant Societies*", Oxford University Press, New York.

Hussi, Pekka Josette, Murphy Ole, Lindberg and Lyle, Brenneman (1993): "*The Development of Cooperatives and other Rural Organizations-The Role of the World Bank*", World Bank Technical Paper No. 199, Africa Technical Department Series, The World Bank, Washington D.C, pp. 1-94.

ILO (1990): "*The Socio-Economic Impact of Technical Cooperation Projects Concerning Rural Development*", ACRD/XI/1990//111, International Labour Office, Advisory Committee on Rural Development, Geneva.

International Labour Conference, 79th Session (1992): "*Democratization and the ILO*", Report of the Director General, Geneva.

Israel, Arturo (1987): *Institutional Development-Incentives to Performance,* The World Bank, Washington D.C, The Johns Hopkins University Press.

Kerala State Planning Board (1996): *Report of the Committee on Reform of Cooperatives in Kerala,* pp. 1-25.

Korten, David (1980): "Community Organization and Rural Development: A Learning Process Approach, "*Public Administration Review*", Vol. 40, No. 5, pp. 480-511.

Krishna, Sudhir (1994): "Restructuring the Institutional Arrangements to Strengthen the Panchayati Raj," In Amitava Mukherjee (ed.) *Decentralisation: Panchayat in the Nineties,* Vikas Publishing House, New Delhi, pp. 189-201.

Linstone, H.A and M. Turoff (1975): *The Delphi Method: Techniques and Applications,* Mass: Addision Wesley, Reading.

Levi, Y and H. Litwin (1986): *Community and Cooperatives in Participatory Development,* Gower London.

Maetz, Materne and Maria, G. Quieti (1987): *Training for Decentralized Planning: Lessons from Experience,* Vol.1, Development Policy Studies and Training Service, Policy Analysis Division, FAO of the UN, Rome, pp. 1-61.

Mawhood, P (1983): *Local Government in the Third World- The Experience of Tropical Africa,* Mawhood, P (ed.), John Wiley and Sons, Chickester.

Nathan, Iben *et. al* (1994): *"Brief Guidelines for the Institutional Screening of Local Governments",* CDR Local Institutions and Rural Development Group, *CDR Working Paper* 94.5, Denmark pp. 1-9.

Nugent, Jeffrey B (1993): "Between State, Markets and Households: A Neoinstitutional Analysis of Local Organisations and Institutions", *World Development,* 21 (4), pp. 623-632.

Olowu, D (1988): *African Local Governments as Instruments of Economic and Social Development,* International Union of Local Authorities, The Hague.

Parker, Andrew N (1995): "Decentralisation: The way forward for Rural Development", *Policy Research Working Paper* 1475, Agriculture and Natural Resources Department, Sector Policy and Water Resources Division, The World Bank.

Prudhomme, Remy (1994): "One the Dangers of Decentralisation", *Policy Research Working Paper* 1252, Transportation, Water and Urban Development Department, The World Bank, pp. 1-35.

Rajaraman, Indira., O.P. Bohra and V.S. Renganathan (1996): "Augmentation of Panchayat Resources", *Economic and Political Weekly*, XXXI (18), May 4, pp. 1071-1083.

Ramachandran, V (1993): *State Legislation on Panchayati Raj Consequent on the Constitution (Seventy-third Amendment) Act, 1992, Issues for Discussion,* State Planning Board, Thiruvananthapuram, Kerala, pp. 1-29.

Rondinelli, D.A (1981): "Government Decentralization in Comparative Perspective: Theory and Practice in Developing Countries", *International Review of Administrative Science,* 47.

Rondinelli, D.A (1983): "Decentralization of Development Administration in East Africa", in Cheema G.S and Rondinelli D.A (eds) *Decentralisation and Development: Policy Implementation in Developing Countries,* Sage, Beverly Hills.

Sankaran, P.N (1994): "Excess Liquidity Syndrome in the Co-operative Banking System of Kerala", *International Congress on Kerala Studies, Abstracts,* Vol.5, AKG Centre for Research and Studies, Thiruvananthapuram, pp. 47-48.

Schonwalder, Gerd (1992): "New Ways of Linking The State and Civil Society: Experiences with Local Government in Lima, Peru", *CDAS Discussion Paper* No. 766, pp. 1-19.

Sengenberger, W (1993): "Local Development and International Economic Cooperation", *International Labour Review,* 132 (3) 315.

State Planning Board (1996): Economic Review.

Streetan, Paul (1991): "Markets and States", Paper prepared for a *Conference on The State and Market in Development,* Brown University, Providence, RI.

Taimni K.K (1993): *Cooperative Development: The Next Phase,* ICA Domus Trust, New Delhi.

Tanzi, Vito (1995): "Fiscal Federalism and Decentralization: A Review of Some Efficiency and Marco Economic Aspects", Paper presented at the *Annual Bank Conference on Development Economics*, Washington D.C, May 1–2.

United Nations (1962): *Decentralization for National and Local Development,* United Nations, Technical Assistance Programme.

Uphoff, Norman (1986): *Local Institutional Development: An Analytical Source Book With Cases,* Kumarian Press, West Hartford, CT.

Uphoff, Norman (1988): "Assisted Self–reliance: Working With, Rather Than For, The Poor", In John P. Lewis [ed] *Strengthening the Poor; What Have We Learned?*, Transaction Books, New Brunswick, N.J.

Uphoff, Norman (1993): "Grassroots Organizations and NGOs in Rural Development: Opportunities with Diminishing States and Expanding Markets", *World Development,* 21 [4], pp. 607–22.

Voluntary Action Network India [VANI] (1995): *State Panchayat Acts–A Critical Review,* VANI, New Delhi.

Wadia, F.K (1996): "Panchayats in India–Approaches and Ground Realities in a Historical Perspective", *Journal of Indian School of Political Economy,* III January–March, pp. 64–97.

Webster, Neil (1990): "Panchayat Raj and the Decentralization of Development Planning in West Bengal–A Case Study", *CDR Project Paper* 90.7, Centre for Development Research, Denmark.

ANNEXURE - 1

Status of Local Government Institutions – Kerala 1996

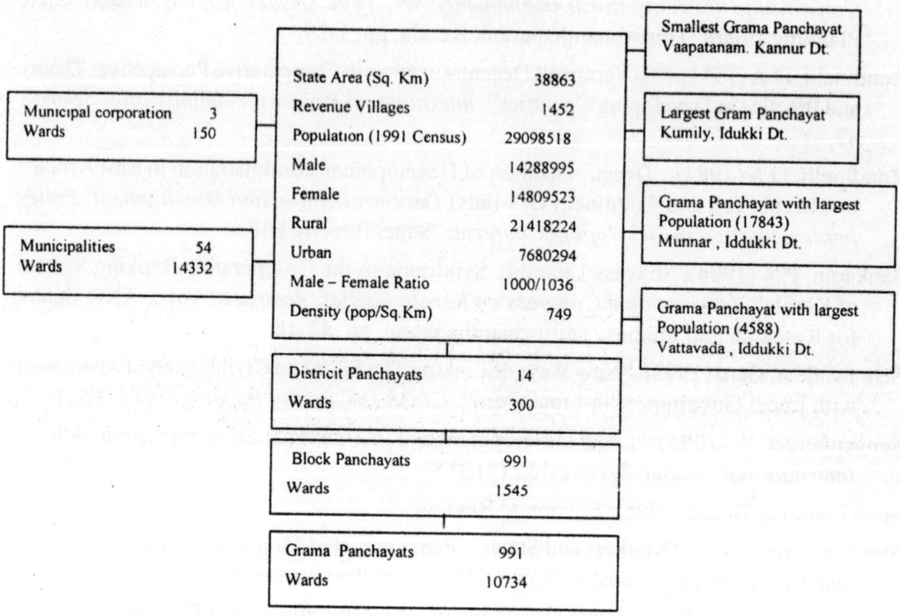

Source: Government of Kerala, Dept. of Panchayats, Panchayat Guide, 1996.

ANNEXURE - 2

Status of Primary Agriculture Credit Societies – Kerala (1992-93)

General Profile	
No. of Villages	1452
No.of Primary Agricultural Credit Societies	1543(65833)
No. Of Panchayats	991
No. of Primary Non-Agricultural Credit Societies	46
Total No. of Cooperative Societies	10275
Total Membership in Credit Societies	10066705(90887000)
Percentage of Borrowning Members to Total Members	41(13)
Deposit Mobilisation (Rs. Lakhs)	33937.70(229400)
Percentage of Viable Societies	92
Percentage of deposits in PACS to total In cooperative banks	72

Deposits (Rs.)		No.of PACS
50-75	lakhs	185
75-100	"	124
1-3	Crores	333
3-5	"	69
5-10	"	27
Above 10	"	7

Per Society Average	
Membership	6371(1380)
Borrowing Members	2587
Share Capital	Rs. 733000(195484)
Deposits	Rs. 9535000(348530)
Working Capital	Rs. 16956000(1915840)
Loans Advanced	Rs. 9484000(819424)

Average Per Member		
Share Capital	Rs.	115(141)
Deposit	Rs.	1496(252)
Loans Advanced	Rs.	1489
Loans Advanced Per Borrowing Members	Rs.	3666(4323)

Source: Handbook on Cooperative Movement in Kerala, 1992-93, Govt of Kerala and Statistical Statements Relating to the Cooperative Movement in India, 1992-93, Part I Credit Societies, NABARD,Bombay.

Figures in parentheses are for the entire country

3

PANCHAYATI RAJ: INSTITUTIONAL DIMENSIONS AND DEFICIENCIES

PANCHAYATI RAJ INSTITUTIONS

Panchayati Raj Institutions (PRIs) in India refer to a statutory, multi-tier administrative structure entrusted with developmental duties and responsibilities by the State legislatures.

Pandit Nehru inaugurated Panchayati Raj at Nagaur in Rajasthan on 2nd October 1959 following the acceptance by the National Development Council of the Balwantrai Mehta Committee recommendation that public participation in community works should be organized through statutory representative bodies. The advocates of the system wanted PRIs to be vibrant institutions of self-government. Panchayati Raj and decentralized planning process have been viewed as part of a "development vision" which conceives organizations such as NGOs and Co-operatives at the grass roots level as instruments of people's participation capable of enhancing democracy, social justice, self-reliance, sustainability and the elimination of exploitation in development programmes.

Despite its intellectual appeal, the new framework for nation building through the village republics failed to take firm roots in the Indian policy. Conceptualization and operationalization are quite different when it comes to issues involving realignment of power hierarchies-political and bureaucratic-devolution of finances and decentralization of development initiatives. From the tale of Panchayats, especially the two where the system was introduced first, one hears reasons like caste politics, limited powers, illiteracy, financial constraints, irregular elections, excessive bureaucratization, etc for their failure. While admitting the above factors behind the saga of Panchayati Raj failure, one has to search for more fundamental drawbacks in the operationalization of the scheme. The road to full-fledged Panchayati Raj, it is feared, will be tortuous, conflict-ridden and long drawn unless the right environment for their healthy survival is guaranteed by the government.

POLICY GAPS IN INSTITUTIONAL DEVELOPMENT

In the evolution of PRIs in India, the role of an institutional setting conducive to their development has received scanty attention from policy-makers. It is true that until recently, mainstream economic theory has, by and large, ignored institutional issues, often stating central propositions with a false air of institutional neutrality. However, in the last few decades institutional weaknesses have been referred to as critical constraints in may programmes for economic development. The World Bank notes that the development community has become increasingly aware of the need for institutional strengthening if development is to be sustained. Hence it is obvious that the success of grassroots democracy and development envisaged under the Panchayati Raj is heavily dependent on the strength and linkages of local institutions, people's participation and empowerment.

The local level institutions involved in development planning are the PRIs at the village, block and district levels, the voluntary agencies, community groups, NGOs, Co-operatives and self-help groups. Unfortunately, even after the Constitution (73rd Amendment) Act, 1992 providing statutory basis to the PRIs, no serious effort has been made towards strengthening the institutional input. Consider, for example, the condition of co-operatives which can serve as the most potent grassroots organization in rural India. Most of them are either under the grip of powerful caste lobbies or under government administration/liquidation with little scope for member participation and education or autonomy in operation. Deficiency in institutional preparedness has by and large plagued the pace of Panchayati Raj in several parts of the country at different stages. Administrators and academics have emphasized this vital institutional environment for the survival and success of Panchayati Raj in India. Promotion of the right type and mix of institutions that interact with the PRIs with necessary incentives and accreditations should have received top priority in the agenda of democratic decentralization and Panchayati Raj.

CONTENT OF INSTITUTIONAL DEVELOPMENT

Institutional development is the creation or reinforcement of the capacity of an organization to generate, allocate and use human and financial resources effectively to attain development objectives. It is the process of establishing or transforming an organization into an integrated and organic part of a community in a way that will help the organization play a proactive role in projecting new values and become an agent of change in the community. A programme for institutional development aims at strengthening, *inter alia,* (1) management systems including monitoring and evaluation, (2) financial management and planning systems, (3) staff development and training, (4) inter-institutional relationships, (5) institutional structures of subsections, and (6) government regulations and procedures. Besides, the socio-political and cultural climate has to be made conducive to the development programmes.

THE CHALLENGES

While decentralized institutions may be more susceptible to local participation and control, by the same token they may be captive to the local structure of power, says Arkadie. The drawbacks of centralized power (insensitivity to local needs and ignorance of local capacities) have their counterparts in the skewed access that can arise from local systems of power prejudice. With more power and resources and less accountability, the Panchayati Raj institutions may become a bed of harassment of the SCs, STs, and women who have no protection. According to Pravin Sheth, funnelling of resources and powers from the Centre on a vast scale to the Panchayats which are not prepared and experienced to use them in proper and competent ways will create massive problems of misuse, wastage and financial mismanagement. In a way, corruption along with power will be decentralized if institutional degeneration is not reversed and supportive institutions not promoted.

FRAMEWORK FOR ACTION

Democracy, service motive, member education and participation, empowerment (especially of the poor) etc constitute the framework for an action plan for institution building at the grassroots level. It is noteworthy that co-operators and promotional agencies like the Co-operative Initiative Panel are campaigning for comprehensive reform of the system in the light of liberalization; however, political will is lacking in the adoption of model co-operative laws envisaging greater autonomy and business-orientation in their operations. Though the role of training has been fully recognized in the scheme of Panchayati Raj, the outflow of trainees from centres like the National Institute of Rural Development is very disappointing. Political and cultural outfits who have a dubious past of being caste/faction–ridden also need reform. The State-PRI relations require appropriate redefinition in the new context.

Another ingredient of institutional reform consists of depoliticization and deofficialization of democratic and service organizations catering to the rural clientele. It seems that the problem of Panchayati Raj is not mainly one of inadequate decentralization of economic planning as it is often made out to be, but of redefining State-PRI relations, reforming the rural socio-political scenario, providing the right type of incentives for institution building and better PRI-collective action sector interface.

CONCLUSION

In the implementation of the Gandhian dream of Panchayati Raj in India, State and Central Governments have to play a promotional role as well, and ensure that the right institutional environment is evolved throughout the country. It is now generally accepted that a critical task of government, neglected in earlier models of economic planning, is to provide the right incentives to stimulate and guide the various actors in the economy. A *corollary of this view is that development policy must focus also on the institutional arrangements that provide the setting in which the incentive system will operate.* Hence it is up to the government, people's organizations, NGOs, and political parties to rededicate themselves and contribute to a supportive institutional setting for the Panchayati Raj, where 'power' must eventually rest with the 'people', both in letter and in spirit as Gandhiji visualized.

NOTE

[Publication of the paper in *Kurukshetra*, Vol. XLV, No. 9, June 1997 is acknowledged].

REFERENCES

Arkadie, B.V (1989): "The Role of Institutions in Development", *Proceedings of the World Bank Annual Conference on Development Economics*, The World Bank.

Bardhan, Pranab (1989): Comment on "The Role of Institutions in Development" by Arkadie, *Proceedings of the World Bank Annual Conference on Development Economics*, The World Bank.

Brett, E.A (1993): "Voluntary Agencies as Development Organizations: Theorising the Problem of Efficiency and Accountability", *Development and Change*, Vol. 24.

Dubey, Ajay (1995): Strengthening Panchayati Raj Institutions: Strategies for NGOs, *Working paper* 87, IRMA.

Gray, C.W et.al (1990): "Institutional Development work in the Bank", Policy, Research and External Affairs, *Working Paper* WPS 437, The World Bank.

Krishna, Sudhir (1994): Restructuring the Institutional Arrangements to Strengthen the Panchayati Raj, in Amitava Mukherjee (ed), *Decentralisation: Panchayats in the* Nineties, Vikas Publishing House.

Mancur, Olson (1996): "Collective Choice Perspectives on Indian Economy", Redistribution as Means of Economic Reforms Mooted, *Financial Express,* January 13,1996.

Mathew, George (1995): A Tale of Two Panchayats, *The Hindu* Daily July 6.

Sheth, P (1991): "Panchayati Raj Revival: Implications and Imperatives" in Hooja and Mathur (ed) *District and Decentralized Planning*.

4

PROLIFERATION AND INTERFACE OF INSTITUTIONS UNDER DECENTRALISED PLANNING: THE KERALA EXPERIENCE

INTRODUCTION

Concern with the institutional dimensions of development is a relatively new phenomenon. The development of local and district level institutional capacity to plan, execute and monitor rural development was neglected and sometimes seriously undermined. The programmes were often designed in a top-down approach within which beneficiaries were not given any authority for decision-making or programme execution (World Bank, 1987). This is clearly reflected in the earlier decentralisation initiatives in Africa and Asia that focused almost entirely on the administrative institutions of government. In the decentralisation experiments of late 1980s and early 1990s, the scope of NGOs, cooperatives and grassroots organisations in the process was still perceived as peripheral. The institutional dimensions of decentralisation have often been given insufficient attention mainly due to the difficulty associated with characterising the role that NGOs and other community-based ogranisations might play in decentralisation. The result is that no framework has been identified that assesses all the dimensions of decentralisation and their linkages (Parker, 1995). Hence the next task of theoreticians and development practitioners of decentralisation lies in the identification of linkages of local governments on the one hand with 'third sector' institutions [like NGOs, cooperatives, voluntary agencies and self-help groups] on the other. Third sector operates between the public and the private sectors. It can be designated variously as the voluntary sector, the membership sector, the self-help sector, the participatory sector or the collective action sector (Uphoff, 1993). In the decision making process, it depends more on voluntaristic accommodation and persuasion with reference to both group and individual interests.

An expanding market [the private sector] and a contracting state leaves many [rural] development issues unattended as the market, by nature, may not be a substitute for the state; the state on the other hand stands reduced in terms of functions and resources. The emerging development challenges could be partially addressed through the promotion of third sector institutions. The third sector has a number of advantages over the state and the market; it consists of voluntary associations, the decision makers are leaders and members, sanctions are driven by social pressure and above all the mode of operation is 'bottom-up'. In fact, the state, as well as the market also depends on the grassroots organisations according to levels of organisation for decision-making and action. When organisations / institutions of different kinds representing all three sectors exist, they can balance and complement one another, the strengths of one sector compensating for the weaknesses of the other (Uphoff, 1993). Those not well served by the private sector can look to government programmes or to self-help organisations. To revive and expand the third sector we require public policies that capitalise on complementary relationships this sector can have with state and market institutions, filling in gaps between the public and private sectors and helping each of them become more effective. What is to mobilised through grassroots organisations or NGOs is not just material contributions and labour but also ideas, management skills and a sense of social responsibility. There is growing evidence that working with and through collective action organisations at local levels can produce improvements in production and well being more far-reaching than the stereotype government-sponsored local development initiatives (Uphoff, 1988). In our programme of [rural] development in the context of decentralisation, the goal should be to combine outside and local funds, personnel and expertise which will also regulate and energise total economic behaviour in better ways than the government and the market can produce independently. This implies explicit recognition and promotion of grassroots organisations and NGOs in all realistic programmes of rural development and community organisation.

INSTITUTIONS

Institutional development is the creation or reinforcement of the capacity of an organisation to generate, allocate and use human and financial resources effectively to attain development objectives. It is the process of establishing or transforming an organisation into an integrated and organic part of a community in a way that will help the organisation play a proactive role in projecting new ideas and become an agent of change in the community (Pareek, 1981). In most of the social science literature, no explicit distinction has been made between 'institutions' and 'organisations'. But there are subtle and important differences between the two. 'Institutions', whether organisations or not, are complexes of norms and behaviours that persist over time by serving collectively valued purposes, while 'organisations', whether institutions or not, are structures of recognised and accepted roles (Uphoff, 1986). Institutions like money, marriage, land tenure etc. are not organisations, while organisations like a local bank, a particular family, a consulting firm etc are not institutions. Institutions like the central bank, the supreme court, the family, the Oxford University, the World Bank etc. are organisations and vice versa.

DECENTRALISATION AND INSTITUTIONS

The local level institutions involved in development planning are the PRIs at the village, block and district levels, the voluntary agencies, community groups, NGOs, cooperatives and self-help groups. Administrators and academics have emphasised this vital institutional environment for the survival and success of Panchayati Raj in India (Issac, Thomas TM and Harilal K.N. 1997, Sankaran P.N, 1996, 1997 and 1998). Promotion of the right type and mix of institutions that interact with the PRIs with necessary incentives and accreditations should receive top priority in the agenda of democratic decentralisation and Panchayati Raj. The state -PRI relations require appropriate redefinition in the new context. Another ingredient of institutional reform consists of depoliticisation and deofficialisation of democratic and service organisations catering to the rural clientele. It seems that the problem of Panchyati Raj is not mainly one of inadequate decentralisation of economic planning as it is often made out to be, but of redefining state-PRI relations, reforming the rural socio-political scenario, providing the right type of incentives for institution-building and better PRI-collective sector interface.

THE KERALA SCENARIO

Kerala has been a pioneer in initiating a number of progressive social, cultural, educational and economic movements and policies over the last few decades. The latest in the series of mass mobilisation for socio-economic change is the 'People's Campaign' for decentralised participatory planning. The primary objective of the campaign has been to empower the local bodies to prepare and implement their own development plans. It envisaged a series of training programmes at various levels as also the formation of a Voluntary Technical Corps to assist the local bodies in plan formulation, technical/financial scrutiny, execution and monitoring. The exercise of Annual Plan [1997-98] formulation was completed in six phases in which the State Planning Board has been functioning as a facilitator to the local governments and the third sector institutions in the process. The People's Campaign for Ninth Plan, according to a report, was transforming itself into a mass mobilisation for genuine decentralisation.

The process of effective decentralisation has opened up a wider space for collective third sector interface in the conceptualisation, formulation, implementation and evaluation of development projects and related interventions in the form of mass mobilisation, conscientisation, beneficiary identification, facilitation of social audit etc. With the launching of People's Planning in Kerala, the existing rural, voluntary and NGOs have diversified their operational frame in the form of increasing involvement and participation in the process. Being a highly literate society with a long tradition of mass mobilisation, institution-building, political and social experiments, Kerala has also witnessed the proliferation and interface of new participatory institutions which have stepped up their scale of involvement in the process of grassroots planning for economic development and social justice. An outline of the institutional proliferation and interface under decentralisation in Kerala is attempted in Chart-1 and 2. The level of operation,

Chart 1
Institutional* Architecture in the wake of Decentralised Planning in Kerala

Governmental Institutions →	Activities	← Institutions (Quasi govtal /NGOs)
	State Level	
Local Administration	Policy Formulation, Administration	
State Development council	Policy Formulation, operational issues	
State Planning Board →	Planning, coordination and Evaluation	← ANERT, EMC, IRTC, COSTFORD, KSSP, Kudumbasree
State Election Commission	Financial Devolution Election	
State Finance Commission	Grievance/Corruption Redressal	
Ombudsman	Training	KILA, SIRD, SCERT, COSTFORD, IMG
	Visual Publicity	C-DIT
	Information Dissemination	Literacy Mission, Information Kerala Mission, KSSP
	Capacity Building	Manaviyam Cultural Mission
	Cultural Input	CapDeck (KILA)
Accountant General, Local Fund Dept, Performance Audit, Finance Dept. (Inspection wing)	Audit	Social Audit
	District	District Planning Committee
	Planning	
	Block Level	VTC (BLEC)
	Project Scrutiny	
	Local Level Panchayat/ Municipality	
Krishi Bhavan, ICDP, PHC, etc	Planning, Evaluation and Monitoring	Task Forces, VTC, NGOs, Cooperatives, CDS, etc.
	Ward Level	
Grama Sabha	Identification of needs, problems and beneficiaries, Audit	ADS, NGOs, Charitable Societies, Voluntary Agencies etc.
	Group Level	
	Participatory processes (leadership, organisation, sharing, services, mobilisation, enterprise etc)	NHGs, NGOs, SHGs, Resident's Associations, Beneficiary groups/Committees.

* Institutions on the anvil (1) Audit commission (2) Appellate Tribunals.

functions and the linkages between government and NGOs are summarised in Chart-1. Institutional and procedural linkages against the background of the environment, facilitation and planning processes are established in Chart-2. The extent of diversification in activities undertaken by the old NGOs and the promotional role played by the emerging institutions have to be scientifically evaluated, documented and disseminated so that it may serve as a guideline for strengthening state-PRI-third sector interface. This paper is a modest attempt in this direction.

Chart 2

Proliferation and Interface of Institutions under Decentralised Planning in Kerala: Environment, Facilitation and Planning

ENVIRONMENT	
73rd and 74th Amendments to the Constitution of India 1992, Kerala Panchayati Raj Act 1994 and Kerala Nagarapalika Act 1994	State Election Commission, Local Self–governing Institutions, State Finance Commission and State Development Council
FACILITATION	
Training	Kerala Institute of Local Administration (KILA), State Institute of Rural Development (SIRD), Institute of Management in Government (IMG), Centre for Development of Imaging Technology (C-DIT), State Council for Educational Research and Training (SCERT), Integrated Rural Technology Centre (IRTC), Centre of Science and Technology for Rural Development (COSTFORD)
Publicity	State Planning Board, Plan Campaign Cell C-DIT
Awareness	Grama Sabha, NHGs, ANERT, EMC, IRTC, COSTFORD, KSSP, NGOs, Literacy Mission, Manaviyam Cultural Mission
Voluntary service/contribution	NGOs, NHGs, Residents' Associations, Charitable Societies, Communal Organisations
PLANNING	
Identification of Projects	Grama Sabha, NHGs, Cooperatives, NGOs, CDS, Residents' Associations etc.
Formulation	DPC, Local Bodies, NHGs, NGOs, Cooperatives, EMC, ANERT etc.
Appraisal	Voluntary Technical Corps, Line Departments etc.
Execution	Local Bodies, Coperatives, NHGs, NGOs, Beneficiary Committees etc.
Monitoring	Local Bodies, NHGs, NGOs, Monitoring Committees etc.
Evaluation	State Planning Board, Local Bodies, NGOs, NHGs etc.
Audit/Transparency	Ombudsman, NGOs, NHGs, Grama Sabha Performance Audit, Social Audit
Capacity Building	Information Kerala Mission, Cap Deck (KILA)

PROLIFERATION AND INTERFACE OF INSTITUTIONS: A DISCUSSION

As already observed, in the wake of decentralised planning in Kerala, a set of new institutions have emerged in the government and non-governmental sectors; the latter having participatory and bottom-up approaches to planning/related interventions. Some of the existing institutions-governmental and non-governmental-have diversified/reoriented their focus and agenda during the People's Campaign for participatory planning. The new linkages forged by the existing and new institutions for greater participation, productivity, transparency, accountability and empowerment, have brought in a number of issues of functional overlapping, lack of integration in plans and procedures of local bodies, horizontally as well as vertically. As governmental institutions are by nature top-down in their approach and are common, to some extent, across states, our discussion on institutional proliferation and interface would be centered mainly on the emerging participatory formations (as highlighted in Chart-3).

Chart 3
Institutions Under Decentralised Planning in Kerala: A Classification

Existing Institutions
Local Administration Department
Kerala State Planning Board
State Institute of Rural Development (SIRD)
Kerala Institute of Local Administration (KILA)
Institute of Management in Govt. (IMG) State Council for Educational Research & Training (SCERT) Centre for Development of Imaging Technology (C-DIT) Centre for Earth Science Studies (CESS) Primary Cooperatives Agency for Nonconventional Energy and Rural Technology (ANERT) Energy Management Centre (EMC) Integrated Rural Technology Centre (IRTC) Centre of Science and Technology for Rural Development (COSTFORD) Literacy Mission Kerala Shasthra Sahitya Parishad (KSSP)
Residents' Associations **Community Development Societies (CDS)** **Area Development Societies (ADS)**

New Institutions
State Election Commission State Finance Commission Grama Sabha District Planning Committee (DPC)
Voluntary Technical Corps (VTC)
Kudumbasree
Neighbourhood Groups (NHGs)
Beneficiary Committees
Performance Audit
Social Audit
Information Kerala Mission (IKM)
State Develpment Council (SDC)
Ombudsman
Manaviyam Cultural Mission

(a) Kerala State Planning Board

According to the framework proposed for the analysis of the dynamics of institutions, we begin with the changing role of existing institutions. Following state policy on decentralisation, 35-40 per cent of plan funds are transferred to local bodies for undertaking locally relevant projects along with mobilisation of additional resources for economic development and social justice. For the first time a women component (roughly of 10 per cent of plan allocation) has been added to the plans of local bodies. In order to supplement the formal efforts of the Board to facilitate the process, a plan campaign cell has been constituted. District planning offices have been strengthened in terms of men and infrastructure, in view of their new tasks in the formulation of district plan and coordination of the activities of local bodies-rural and urban. As facilitators of the process, the Board has been actively associated with activities like training, publicity, awareness creation, surveys and studies, state level coordination of evaluation, monitoring etc.

(b) Plan Campaign Cell

People's Campaign programme for Ninth Five Year Plan, as already stated, is an innovative step in the planning horizon in Kerala. Considering the physical limitation of the existing infrastructure of the Planning Board to engineer the peoples plan programme, a Plan Cell was conceived and it started functioning in the State Planning Board premises. The cell is expected to supplement the activities of the SPB in connection with the planning programme by making available suitable human resources, which are drawn from various official boards, voluntary agencies and NGOs. Though it is not a governmental body, it is expected to give advice and provide necessary human resource and skill for supplementing the functions of SPB. By collecting and disseminating information pertaining to peoples plan from every nook and corner of the state, the cell is discharging the function of an informal institution, though it is not originally meant for this.

(c) Kerala Institute of Local Administration (KILA)

The institute has been identified as the designated lead organisation to implement a project 'Capacity Development for Decentralisation in Kerala (Cap Deck) with Swiss Agency for Development and Cooperation grant of Rs.1054 lakhs over a period of three years. The project aims to support the democratic decentralisation process by contributing to the conceptualisation, development and implementation of a capacity building strategy and programme that will strengthen local participation and enable rural and urban local bodies to effectively perform their duties and responsibilities (GOK, Planning Board, 1999). Besides the above project, the institution has been arranging training programmes for officials and non-officials of local bodies, over the years.

(d) Residents' Associations (RAs)

Residents' Associations in urban areas of the state have been a peoples organisations registered under societies registration act to inculcate feeling of unity in neighbourhood and promote mutual help. The life style, standard of living and diversity of activities have combined together to offer an entirely different ambience to RAs, when compared to rural areas. For example, while the formation of grama sabha was a success in rural areas, by some reason or other it was not a success in urban setting. At the same time RAs were found to be efficient organisation to discharge the activities performed by the grama sabha in rural setting. Identifying the potentiality, today RAs are assigned more duties and are actively participating in the planning process. In view of the higher educational and technical background of the urban citizens it is found that much time and cost can be saved in identifying and implementing programmes with the participation of RAs. The role of RAs are increasingly acknowledged and new ones are formed for spreading and implementing programmes offered through decentralised set up.

(e) Community Development Societies (CDS) and Kudumbasree

The history of CDS may be traced to participatory poverty eradication programme for women launched in Alleppey Municipality as part of the centrally sponsored UBSP. The Alleppey model was gradually extended to all municipalities and 96 grama panchayats in Malappuram District. It is a good example of participatory development/poverty eradication intervention by a group process of 15 to 40 women each identified from poor families by the households based on poverty index consisting of 9 risk factors. CDS is also an institutional arrangement for eradication of poverty through participatory thrift and credit schemes; it is not subsidy-based but self help-based. Based on savings mobilised through weekly meets, CDS could facilitate the flow of institutional credit to its members on group guarantee terms for undertaking income generating activities/micro enterprises. As the repayment rate has been as high as 95%, refinance has also been ensured from NABARD.

Advent of decentralised planning facilitated the graduation of CDS model into a state level project 'Kudumbasree' in 1997 under the Kerala Poverty Eradication Mission. Section 284 of Kerala Nagarapalika Act 1994 and Schedule I Annexure 4, Function No. 11, promote CDS as constitutionally valid registered societies. The objectives of "Kudumbasree' are:

- Promotion of productivity, efficiency and capability
- Development of an informal bank through thrift and credit schemes
- Improvement of income through self-employment ventures
- Health promotion and development of basic facilities
- Assured education to children in risk prone households, and
- Linking of the poor with democratic decentralisation.

The organisational pattern of NHGs is adopted for CDS and they are managed by committee of 5 for a period of 2 years. Household level NHGs (all households) are further linked to ward level Area Development Societies (ADS) and to Panchayat/Municipal level CDS of exclusively poor households. The byelaws linking CDS with local bodies have been approved by the Government. Ward level coordination committee consists of a convenor (ADS Chairperson), and other members viz Presidents of NHGs, community organiser, local volunteers, auxiliary nurse, midwife, anganwadi workers and literacy workers. At the Panchayat/Municipal/Corporation levels also such coordination committees are established. The concept of 'linkage banking' introduced by NABARD in 1991-92, the Micro

Credit Summit 1997 Washington, the involvement of Local Administration Department, line departments, the HUDCO, UNICEF, Commercial banks, the intervention of CDS members through 'quality circles' etc have made the CDS one of the most effective participatory mechanism in decentralisation. It is a unique model of women empowerment (socially, economically and politically) that has invited in UN attention; Alleppey CDS has won the UN 'We the People's Award' in 1995, and a former chairperson and later president of Alleppey ADS/CDS was elected to the Municipal council. The CDS model deserves promotion as its approach is participatory identification of the poor based on non-income criteria the emphasis is on self-help through thrift and credit (and not subsidy) and as group guarantee pays dividends in terms of high repayment/effective use of credit.

We have seen that membership/leadership in grassroots institutions like CDS has promoted political empowerment of women. Similarly, a recent study has indicated that participation in Peoples Planning has empowered women representatives of local bodies in terms of awareness, and capacity to manage staff/meetings, interact with the public and improve the art of public speech (See Table 1).

(f) Voluntary Technical Corps (VTC)

Voluntary Technical Corps is a link between the bureaucracy and the people that acts as a catalyst/lubricant of local level planning through the mobilisation of concealed capacity and expertise of retired men/women for socially important productive purposes. The main function of the VTC at Panchayat, Block and District levels is technical scrutiny of projects of local governments for consideration and approval by the District Planning Committee. It also helps local bodies in the formulation of financially and technically viable projects. The advantage of the VTC is that many local members of the body render their services free of cost and with a sense of participation.

(g) Neighbourhood Groups (NHGs) and Self-Help Groups (SHGs)

In the decentralisation experience of Kerala, the role of NHGs and SHGs as grassroots organisations have become unique (See Chart 4). NHGs are nonstatutory and participatory groups of 50 to 70 families in a compact area. Whereas the 'Grama Sabha' is a creation of the constitution, NHGs are creations of local bodies to function as parallel institutions that can compensate for the limitations of the former (Issac Thomas T.M. 2000). It is hoped that participation in grama sabha can be improved through NHGs. The context for the formation of NHGs has been the success stories of CBNP/CDS/SHGs, Socio-Economic Surveys, Projects (drinking water, sanitation etc.), study/research undertaken by NGOs, voluntary agencies etc. The popular functions of NHGs are (1) formulation of schemes, projects, execution, evaluation and monitoring (2) identification of beneficiaries and (3) additional resource mobilisation. They are associated with projects of sanitation, drinking water, anganwadis, nutrition, school education, thrift etc. A number of production units are managed by them; they also arrange medical camps, sports

Table 1
Empowerment of Women as Local Body Representatives: Impact of Participation in People's Planning

Impact of Participation	*Awareness (%)*				*Capability (%)*							
	PR Act/ Rules		*Local Planning*		*Management of PR staff*		*Management of council other meetings*		*Interaction with public*		*Public speech*	
	1995	*1998*	*1995*	*1998*	*1995*	*1998*	*1995*	*1998*	*1995*	*1998*	*1995*	*1998*
1. Very Poor	30.72	0.27	30.29	0.00	39.21	2.58	19.34	0.13	4.64	0.00	27.44	0.60
2. Poor	42.86	10.46	32.30	2.95	29.03	14.55	32.42	4.64	19.09	0.47	42.73	12.10
3. Satisfactory	18.91	44.80	18.60	23.04	22.67	35.68	31.87	31.09	31.05	12.15	23.65	51.20
4. Good	6.30	32.18	15.52	43.72	6.74	31.21	11.32	38.89	23.86	32.00	4.72	25.0
5. Very good	1.21	12.29	3.29	30.29	2.35	15.98	5.05	25.25	21.36	55.38	1.46	11.10
Total	100.00	100.00	100.00	100.00	100.00	100.00	100.00	100.00	100.00	100.00	100.00	100.00

Source: Women Representation Survey, 1998 as reported in Equality in Development (2000), Consolidated Report of the Papers presented in the International Conference on Democratic Decentralisation, People's Plan Campaign, SPB (ed) T.N Seema and others, p. 22.

Chart 4

Grassroots Organizations under Decentralized Planning: The Emerging Scenario

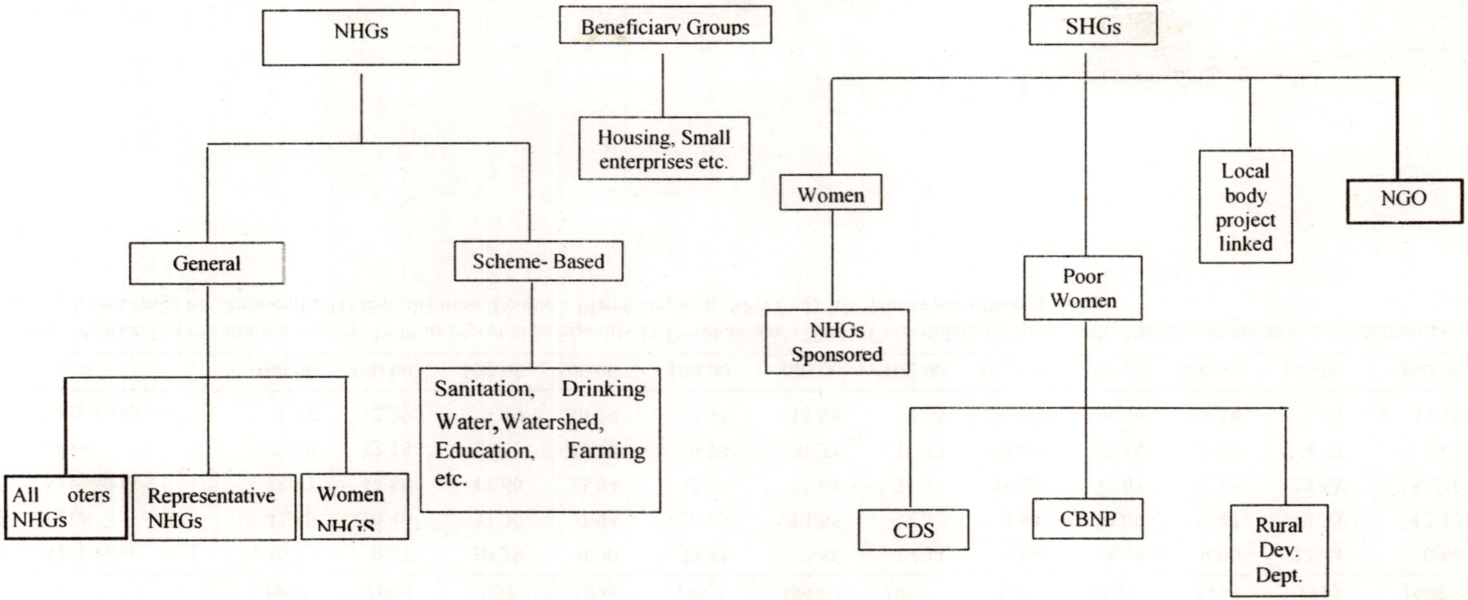

Source: Issac Thomas T.M. (1999) Peoples Planning and NHGs: Experience and Lessons, in TN Seema etal (ed) Ayalkootta Sangamom, 1999 (Malayalam), State Planning Board 1999, P.xxi.

meets, marketing of crops, cultural processions, classes to improve pass rate of school children in risk-prone households, prevent dropout in schools etc. NHGs have exhibited highly effective interventions in beneficiary identification and formulation of beneficiary groups. They also play active role in services, awareness, settlement of disputes as Nyaya Panchayats, women empowerment, collection of socio-economic information, social auditing and monitoring of projects. The growth of NHGs in 96 selected Panchayats has been very impressive; from 9 in 1996 the number has grown up to 30 in 1997, 55 in 1998 and 223 in 1999 (Issac, Thomas T.M. 1999). According to a rough estimate made in 1999, NHGs operates in 223 Grama Panchayats and they are more popular in North Kerala. We get a picture of the diverse interventions of NHGs from Table 2.

Table 2
Functions Performed by NHGs in Planning/Development of 96 Selected Panchayats

	Activity	*No. of Local Bodies*	%
1.	Project discussion	57	59
2.	Project execution	72	75
3.	Identification of beneficiaries	75	78
4.	Beneficiary groups formation	13	14
5.	Saving schemes	30	31
6.	Settlement of family disputes	42	44
7.	Educational programmes	42	44
8.	Health services	52	54
9.	Cultural functions	48	50
10.	Children's forum	11	12
11.	Women forum	35	36

Source: Issac, Thomas. T.M. (1999) p. xiii.

We have a three-tier NHG structure in Kerala consisting of NHGs at household level, Development Committee at ward level (serving as an informal executive of grama sabha) and Development Committee at Panchayat level. There are two types of Development Committees, first consisting of representatives of tier 2, few officials and prominent local persons and the second comprising of representatives of all NHGs. It is the general council with an elected executive. The popular types of NHGs (as may be seen from Chart 4) are all voters NHG of 40-50 families, representative NHG (one male and female voters each from a neigbourhood of 40-50 households) and women-only NHGs. An idea of household coverage of NHGs in selected 96 Panchayats is conveyed in Table 3.

Elected Chairman, Convenor and few members including women manage the NHGs. In a number of Panchayats there exists 'Panchayat Vikasana Samithi' (PVS) in which all the grassroots organisations are represented. The president of the Grama Panchayat acts as the President of the PVS and the other elected representatives as ex-officio members. As the PVS is a registered body under the Charitable Societies Act, it can seek non-plan funds as an NGO. It uses NHGs as a link between local body and the people.

Table 3
Household Coverage of NHGs in 96 Selected Panchayats

Participation rate (%)	No. of Panchayats Based on membership	Based on members attendance
Below 25	32	27
26-50	12	26
51-75	14	31
76-99	24	12
100	14	0
Total	96	96

Source: Issac, Thomas T.M. 1999, p. xvi.

SHGs are institutional agencies created for (1) women empowerment by NHGs, (2) better execution of local projects, (3) eradication of poverty of urban women (under agencies like CDS, CBNP etc.) and (4) as part of capacity building by NGOs. As observed in the context of CDS, women SHGs, based on saving schemes facilitate flow of plan funds and bank credit to members on mutual guarantee terms. Integrated central government scheme for poverty eradication 'Swaran Jayanthi Grama Rozgar Yojna' has visualised SHG participation in its implementation. Another source of favourable environment for the growth of SHGs in the state has been the decision to allocate 10% of plan funds for women empowerment/component plan. State policy has gone a long way in the promotion of local body sponsored SHGs, using plan funds set apart for their formation. Few SHGs have also comeup following the promotional role played by NABARD, Cooperatives, NGOs etc. SHGs as a whole play a supportive role in local-level planning, empowerment and participation. However, cases of SHG duplication has also been noted as part of the attempts to avail plan funds through the floatation of new SHGs avoiding the existing institutional capacity. Popularity of micro enterprises/credit programmes has also attracted the formation of new SHGs at Grama Panchayat level, for provision of training at Block Panchayat level and for marketing services at the Block/District Panchayat levels.

(h) Beneficiary Committees

Implementation of public works by any genuine beneficiary committee or public committee was very difficult in the administrative setup that existed. In order to render transparency and to maximise benefits from public works, the involvement of 'beneficiary committees' replacing contract system has been emphasised. In the year 1997-98, a system of local rates were fixed by district level expert committee. It was discontinued in the next year and an excess limit was fixed in relation to PWD rates. Any excess expenditure over the outer limit would have to be raised by the local body or the beneficiary committee in the form of local contribution in cash or kind. It was expected that genuine and efficient beneficiary committees would be able to execute the work at much below the legally permitted excess limit. Block/Municipal and District level expert committees constituted for project appraisal of local bodies were endowed with powers of technical sanction of projects replacing the department officials. In a significant departure from the past, permission is granted to give mobilisation advance to the beneficiary committee. Besides reducing cost of projects, the system of beneficiary committees facilitated local resource mobilisation and transparency in the execution.

(i) Performance Audit

Performance audit is a nontraditional approach to audit (the traditional type of audit of local bodies is done by the Accountant General, Local Funds Dept, and Finance Dept-Inspection wing). It promotes the right to information involving concurrent evaluation of projects. Rather than a mechanism, to bring to book persons responsible for lapses, it aims at improving project performance/ transparency through frequent interventions. Performance audit also involves spot verification, inspection and interaction with project beneficiaries. Its operational frame consists of evaluation of assets/services in terms of quality, sustainability, timeframe of execution, access of the people/peoples' organisations to the planning process etc (International Conference on Democratic Decentralisation, 2000) The advantages of the process are:

- Quarterly periodicity of the exercise and notification by the regional/district auditing officer
- Opportunity available to the public to bring to official notice complaints, if any
- Circulation of questionnaire by auditing authority to the Panchayat to collect information
- Field visit for spot evaluation of projects and collection of opinion / complaints from the people
- Reporting of audit observations within a week and discussion thereof in the meeting of the Panchayat/Grama Sabha on the questions raised and the reply/clarifications submitted, and
- Interaction with audit officers on issues found unacceptable to the local body for mutually agreeable resolution.

Though performance audit is a welcome step in the interest of transparency and the right to information, it is not clear as to what type of audit will prevail ultimately if there is a conflict in the different systems of audit.

(j) Social Audit

Social audit is a participatory mechanism devised by local bodies in their own way; there is no uniformity in the methodology. Hence we propose to look at social audit with reference to a case study of Nellanad Grama Panchayat (Transparency: New Perspectives, ICDD 2000). As a prelude to local audit, it was decided to undertake all the public works through beneficiary committees. The first audit was carried out in an open assembly consisting of 20 and 60 elected members from the ward level NHG and Grama Sabha respectively, members of panchayat level NHG health promotion samithy, implementing officers and elected representatives. The project components subjected to social audit were:

1. Quality, fund utilisation and implementation pattern of construction activities
2. Working of beneficiary committees
3. Time-frame of implementation
4. People's mobilization and additional resource mobilisation
5. Promotion of peoples' right to information
6. Effectiveness of NHGs
7. Interaction between people, elected representatives and officials, and
8. Functioning of VTC.

The methodology adopted in Nellanad had the following steps:

- Formation of an internal audit samithi as recommended by the Grama Sabha, consisting of technical experts, professional auditors, people of high social standing /acceptability, social activists etc
- Collection of information, documents, reports etc. from Panchayat bodies, implementing officers, beneficiary committees and related institutions by the Samithi
- Preparation of an audit report covering general administration, development projects, services, operation of popular committees etc
- Circulation of the report to NHGs for comments
- Consideration of the report by the Panchayat, Grama Sabha and Development Seminar, and
- Finalisation of the report and onward transmission for performance audit and local fund audit.

The social audit report covered individual beneficiary schemes, community assets, public institutions and social service organisations. As in the case of performance audit no system has been devised for the resolution of possible conflict that may arise from social and other systems of audit. If social audit is subjected to official modification, its participatory colour is lost.

(k) Information Kerala Mission (IKM)

The IKM inaugurated in August 1999 was to be launched formally in January 2001. It is a sequel to the People's Plan Campaign (PPC) and resorts to a very novel methodology of participative electronic governance and human-centered informatics system building. Its methodology grew from the Kumarakom experiment launched as part of the 'Information Kerala Project' in 1998. The mission seeks to eradicate absolute poverty in ten years through concerted community action under the leadership of local governments, by facilitating organisations of the poor for combining self help with demand led convergence of available services and resources to tackle the multiple dimensions and manifestations of poverty holistically. The first phase of the mission involves establishing connectivity between the State Planning Board, the District Planning Offices and 1215 local bodies, developing a series of software applications aimed at making all the aspects of the local body administration and decentralised plan implementation programme efficient, responsive and transparent. The second stage would involve integration of the databases with other community databases and offering value-added services centred on the integrated databases and locally generated content.

Though information is a vital input in planning, it has to be generated and disseminated as cheaply as possible to be used by local bodies and the public. More emphasis need to be laid on strengthening local production capacity, investment opportunities, marketing etc. which are the weak links in the Kerala model of development. Information delivery at grassroots must facilitate reversal of the stagnation in production sectors.

(l) State Development Council (SDC)

The SDC is headed by the Chief Minister and consists of the entire cabinet, Leader of Opposition, Vice-Chairman of the State Planning Board, the Chief Secretary, all the District Panchayat Presidents/Chairpersons of District Planning Committees and representatives of other tiers of local self government. Modeled on the NDC, this institution is expected to take the lead in policy formulation and in sorting out operational issues.

(m) Ombudsman

It is a high power institution consisting of High Court and District Judges, secretaries to Government and eminent public-men selected in consultation with the Leader of the opposition. It has been given vast powers to check malfeasance in local self governments in the discharge of developmental functions.

(n) Manaviyam Cultural Mission (November 1, 1999-January 26, 2001)

It is a cultural project being implemented through the local bodies in the state. The mission aims at documentation of Panchayat level (heritage) history, art and cultural forms and bring out volumes on 'Panchayat Vijnaneeyam' by the Kerala Gazetteers Department. State Archaeology Department has been identified for the establishment of cultural heritage centres in Panchayats. Under the mission, it is proposed to construct millennium monuments by District Panchayats, undertake status study on women, organise cultural festivals, exhibitions etc in order to provide cultural support to the decentralisation programme.

CONCLUSIONS

Though grassroots and other institutions have contributed to facilitation, participation, empowerment, transparency, resource mobilisation, efficiency and productivity in decentralised planning, a number of issues remain to be looked into. Are we fully utilising the institutional capacity bequeathed by the era of top-down planning? Isn't it necessary to prevent duplication of institutional capacity?

Can the growth of NHGs overshadowing Grama Sabha be considered a healthy development? If NHG proliferation is due to the weaknesses of Grama Sabha, is it meaningful to strengthen it? To what extent institutional involvement has facilitated additional resource mobilisation? How far the old/new institutions have become sustainable? What is the extent of democracy, participation, empowerment, transparency etc. that prevails in the institutions, in terms of women/SC/ST, other backward class and minority representation/leadership?

Despite the claim that decentralisation do not seek to create any parallel structure but seeks to restructure the existing ones (Issac and Harilal, 1997), a number of new institutions have come into being like the NHGs under direct local body sponsorship. Information Kerala Mission is another innovation. It is very much necessary to look into the status and effectiveness of 'Grama Sabha' as an assembly of voters and the circumstances under which it is being rendered redundant as in Kerala. Do we have a 'Kerala Model of Grama Sabha' evolving through the proliferation and interface of institutions under decentralised planning? Does the emerging scenario of participatory bottom-up planning and the institutional processes provide a turning point towards sustaining the Kerala model of development characterised by high level of social development and weak, if not stagnant, production sectors?

Local level development has the potential of tapping indigenous knowledge systems of farmers, tribals, artisans etc. Despite high rate of institutional participation, it is doubtful whether such knowledge sources have been adequately mobilised. Though preliminary reports indicate that decentralised planning and institutional participation have improved the rate of local resource mobilisation, a detailed examination of the question is overdue. In the light of the analysis of the role of institutions in the process of decentralisation, we maintain that the dynamics of institutional intervention, particularly its implications for state-PRI-third sector interface need to be evaluated at the earliest.

The Kerala experience of participatory development is in its infancy. Its theoretical framework and operational modalities have to be strengthened through evaluation and case studies of successful projects/schemes at Panchayat, Block and District levels. Though the time is not ripe enough to make an exhaustive evaluation of the functioning of the system, the Kerala experience is an indicator as to how processes can develop. It is quite natural that new institutions/organisations crop up in participatory development. But it will be more appropriate to make an assessment of the various agencies/institutions and their field of operation so that their present status can be made more explicit. This will enable us to assign new and additional duties and responsibilities to them so that unsustainable and competitive proliferation and interface can be minimised once the system gets stabilised. It would also be meaningful to look at local specificity, if any, in the proliferation and interface of institutions. These will save resources and pave the way for the introduction of new institutions with innovative ideas.

NOTE

[Publication of the paper in S.P. Jain (ed), 'Emerging Institutions for Rural Development', 2001,Vol. 1, NIRD, Hyderabad, is acknowledged].

REFERENCES

Government of Kerala, State Planning Board (1999): Economic Review, 1999.

International Conference on Democratic Decentralisation (2000): *Transparency: New Perspectives* (Malayalam) N. Jagajeevan et. al [ed] collection of papers, p. 250-260, 304.

Issac, Thomas T.M and Harilal K.N (1997): "Planning for Empowerment: People's Campaign for Decentralised Planning in Kerala", *Economic and Political Weekly*, Vol. xxxii, Nos. 1 and 2, January, p. 4-11.

Issac, Thomas T.M. (1999): "People's Planning and NHGs; Experience and Lessons" (Malayalam), in T.N. Seema et. al (ed) Ayalkootta Sangamom, State Planning Board, p. xi.

Kudumbasree, Thiruvananthapuram District (1999): Community Development Society Handbook [Malayalam].

Pareek, Udai (1981): *Beyond Management*, New Delhi, Oxford and IBH.

Parker, Andrew N (1995): "Decentralisation-The Way Forward for Rural Development?", *Policy Research Working Paper* 1475, The World Bank, Agriculture and Natural Resource Department, Sector Policy and Water Resources Division, p. 23.

Sankaran, P.N (1996): "Institutional Reforms for Panchayati Raj: Towards Linking Cooperatives with Local Goverments", in R. Rajagopalan (ed) *Rediscovering Cooperation*, Vol. I, Bases of Cooperation, Institute of Rural Management, Anand, Gujarat.

Sankaran P.N (1997): "Panchayati Raj: Institutional Dimensions and Deficiencies", *Kurukshetra*, Vol. XLV, No. 9, June.

Sankaran P.N (1998): "Decentralised Planning: Opportunities for NGOs", *Employment News*, Vol. xxiii, No. 28, 10-16 October.

Uphoff, Norman (1986): *Local Institutional Development: An Analytical Source book, with cases,* West Hart ford, CT: Kumarian Press.

Uphoff, Norman (1988): "Assisted Self-Reliance: Working with, rather than for, the poor", in John P. Lewis (ed), *Strengthening the Poor: What Have We Learned?,* New Brunswick. NJ: Transaction Books.

Uphoff, Norman (1993): "Grassroots Organisations and NGOs in Rural Development: Opportunities with Diminishing States and Expanding Markets", *World Development*, Vol. 21, No. 4, p. 609.

World Bank (1987): *World Bank Experience with Rural Development*, Operations Evaluation Department, Report No. 6883, Washington D.C: World Bank.

Part - III

DECENTRALISATION, INCLUSION AND PARTICIPATORY JUSTICE

5

TOWARDS AN ARTISAN PLAN

INTRODUCTION

In many developing regions, despite the market regime/ process of economic planning, there exists some segments of the population which are not covered satisfactorily by either of the two processes. Lack of promotional institutions, weak educational capability, backward technology, poor access to investment resources etc. tends to perpetuate their marginalization. Besides being highly unorganized, scattered, stratifie and fragmented into splinter groups, these categories are also not effectively covered by social security and insurance network. The plight of traditional artisans of India in general and of Kerala in particular comes closer to the scenario of marginalization depicted above.

According to the National Productivity Council, an artisan is a self account worker or a wage earner engaged in the manufacture of tangible products largely through the application of his skills which were acquired either traditionally or through formal training and which are not personalized services of any nature (NPC, 1979).

The Committee for Artisan Development (1979), Govt. of Kerala in its definition of artisan combines the NPC and Reserve Bank of India (All India Debt and Investment Survey 1971–72) approach. The Committee observes: "an artisan is an own account skilled worker or a skilled wage earner, who has acquired his professional skill either by formal training or by tradition and is engaged in the production of tangible goods and or productive services in tiny establishments using low cost or intermediate level of technology". National Sample Survey Organisation considers artisans as own account skilled workers and handicraftsmen engaged in occupations like spinning, weaving and tailoring and shoe makers, carpenters, stone cutters, carvers, blacksmiths, sheet metal workers, jewellery works, makers of musical instruments, brick layers etc (NSS 48th Round, 1992).

ARTISANS OF KERALA

The different categories of artisans in the state identified by the Committee for Artisans Development (1979) are the following: (1) Wood workers–Carpenters

(2) Gold/Silver metal workers–Goldsmiths (3) Iron/Steel metal workers–Blacksmiths (4) Copper/ Bronze metal workers–Copper / Bronze smiths (5) Stone Masons and Brick layers etc.–Masons (6) Earthenware vessel makers–Potters (7) Shoe makers–Cobblers (8) Handloom weavers–Weavers (9) Fiber/Screw Pine/ Rattan workers–Handicraftsmen and (10) Ivory/Wood carvers–Handicraftsmen. Among the artisans of Kerala, the prominent groups are Carpenters, Blacksmiths, Goldsmiths, and Handloom weavers. In Kerala, as in many other parts of the country, the artisans are a socially and economically depressed lot. They do not belong to the ownership class to supplement their income from work through agricultural production. The economic status of traditional artisans performing activities like carpentry, blacksmithy, goldsmithy, masonry etc is determined by the scale and earnings from work participation, size of industrial and household asset holdings, level of technology assimilation, magnitude and effectiveness of official development initiatives etc. Educational backwardness, poor industrial and household asset base, raw material shortage, competition from the factory system etc have precipitated instability in their employment and earnings over time. Marginalization of these classes from the mainstream economic activity and the largescale eviction / displacement of the most indigent artisans from traditional occupations have exposed them to the risks of changing technology, emerging market scenario, high cost of input, falling income, unstable employment and poverty (Sankaran, 1999). State initiative towards the development of artisans (through the Department of Industries and Commerce, Khadi and Village Industries Board, State Handicrafts Development Corporation, Artisans Development Corporation, Industrial co-operatives etc) was conceived against the above setting.

However, a number of studies have found that all the development efforts in the past have bypassed the artisans, particularly the indigent among them (Committee for Artisan development 1979, Sankaran, 1999). Social security and development interventions introduced so far have touched only the tip of the iceberg.

PEOPLE's PLANNING AND TRADITIONAL ARTISANS

One of the crucial gaps in artisan development is the absence of a detailed socio economic database and status report on their welfare. The Commencement of local-level decentralized planning has brought in the right environment for a multipronged strategy towards artisan rehabilitation.

The primary objective of the People's Campaign for Ninth Plan in Kerala has been to empower the local bodies to prepare and implement their own development plan (Issac and Harilal , 1997). Decentralized planning also provides opportunities for establishing linkages with grassroot organizations like cooperatives, NGOs and other agencies in the third sector (Sankaran 1996, 1997 and 1998). It is widely acknowledged that decentralized planning process commencing from the gramasabha has ensured the informed participation of all sections of the society especially the weaker groups. However, a close look at the scenario indicates that there are some occupational groups that are left out from the ambit of the planning process. One such category is traditional village artisans scattered all over the state. For majority of the artisans grappling with the sinking traditional occupations, there is neither uniformity in wages and adequate social security, nor facilities for upgradation of skills, technology etc. Left to themselves, they are not likely to articulate their development and rehabilitation in terms of local–level and skill–specific plans. Though artisans are found in every village, they are much scattered and cannot be organized into effective groups in grama sabha. The development projects, that we have already in our shelves does not suit the requirements of the artisans. Because of poor funding and top–down style of operation, the existing corporations and schemes (said to have been instituted for the benefit of the village artisans) have also failed to address the basic issues. It is specifically against the above background that local level innovative schemes and artisan plan need immediate attention.

OBJECTIVES OF THE STUDY

The following objectives have been framed for the study:

1. To undertake a quick review of the schemes/projects formulated for the benefit of (traditional) artisans by the local bodies in the People's Plan, and
2. To formulate a participatory 'artisan plan' for traditional artisans–carpenters, blacksmiths and goldsmiths–on the model of Special Component Plan/Tribal Sub Plan for implementation by the local bodies.

METHODOLOGY

The study has been carried out in two stages, viz the evaluation of artisan projects/schemes under Annual Plans 1997-98, 1998-99 and formulation of a model artisan plan. For the evaluation part, the Development Reports of all the Grama Panchayaths, Municipalities and the Corporation in Thiruvananthapuram District were examined to gather basic information on artisan elusters and activities. An evaluation of the specifie schemes involving artisan activities of selected 15-20 Grama Panchayaths in the district constituted the second phase of the evaluation part. Selection of the Panchayaths mainly depended on the information available from Development Reports relating to artisan clusters and artisan –related projects. The role and effectiveness of artisan–related projects were assessed in terms of physical and financial progress. Secondary sources of information like the statements of projects of Grama Panchayaths, Development Report (1997-2001), Economic Review etc. were used for the evaluation exercise.

For the development of a participatory artisan plan, which constituted the second stage of our study, we had to select a Panchayath and a ward with high artisan concentration. The scanning of the Development Report and preliminary interaction with artisan organizations convinced us that Soolamkudy, Ward IX of Poovar Grama Panchayath would be a suitable area for the formulation of an artisan plan using the participatory approach. A development seminar and Participatory Rural Appraisal (PRA) workshop were organised in Soolamkudy. In the development seminar, the rationale, objectives and methodology of the People's

Plan were discussed; development issues of traditional artisans were also posed during the deliberations. The PRA technique, developed by Robert Chambers, allows rural people to analyse, plan and to act so as to bring about sustainable improvement in the condition of their living. One of the widely appreciated merits of the approach of PRA is that it ensures the maximum participation of the stakeholders, allows people to be the decision makers and keeps the outsiders/ experts simply as facilitators. In order to collect basic socio-economic information about participating artisans, a structured questionnaire was also circulated in the seminar session. Thus the methodology of the present study consists of the survey technique for collection of basic socio–economic information and the participatory approach for the formulation of artisan plan.

The workshop session was used to develop an artisan plan through Force Field Analysis (FFA) of PRA. With increasing emphasis on decentralization and participatory approaches to development, there is a tremendous scope for the application of FFA in PRA. According to Kurt Lewin, any situation can be viewed as a state of temporary equilibrium caused by two sets of opposing forces namely driving/facilitating forces and restraining/resisting forces. All the forces are assigned weightages according to their perceived effect on the problem situation. FFA makes it easy thereby, to pinpoint the forces, which need to be further strengthened and the ones which need to be weakened. Drawing on visual depiction, FFA provides people opportunities to think of forces affecting the problem in question. Though individuals can conduct FFA, it is more effective in small groups. The technique therefore fits well with the basic tenets of PRA where the methods are supposed to enable local people to share, enhance and analyse their knowledge of life and conditions, to plan and act (Someshkumar, 1999).

After initial rapport building and explaining the objectives of the exercise, the problem would be written down in a sheet of paper in words/figures/diagram/ symbol. Participants would be asked to identify and represent the driving and resisting forces in terms of a diagram. Considering the present status as a horizontal line, the driving forces and resisting forces are drawn on it as vertical lines in the upward and downward direction respectively (See Diagram 1).

The participants can assign weight to the forces in terms of their length (in each direction) from the horizontal equilibrium point. Once they are satisfied with the diagram and relative strength of the forces, the next step is to ask them to discuss how they can bring about change in the situation; which of the drawing forces can be reinforced and which resisting forces can be diminished. The final picture/position that emerges would be drawn as a diagram for further action. For the assignment of weights to the forces, scoring, ranking and frequency methods are also available. In the present study, we made use of the ranking method.

PEOPLE'S PLAN AND ARTISAN DEVELOPMENT: A REVIEW

In our attempt to develop an Artisan Plan, both secondary and primary sources of data have been used. As an understanding of the artisan clusters and nature of projects (already taken up for implementation or fully executed) was necessary before we move towards developing a model artisan plan, we undertook a quick review of the Development Reports of all the Grama Panchayats in Thiruvananthapuram District (See Chart 1).

Diagram 1: Force Field Analysis

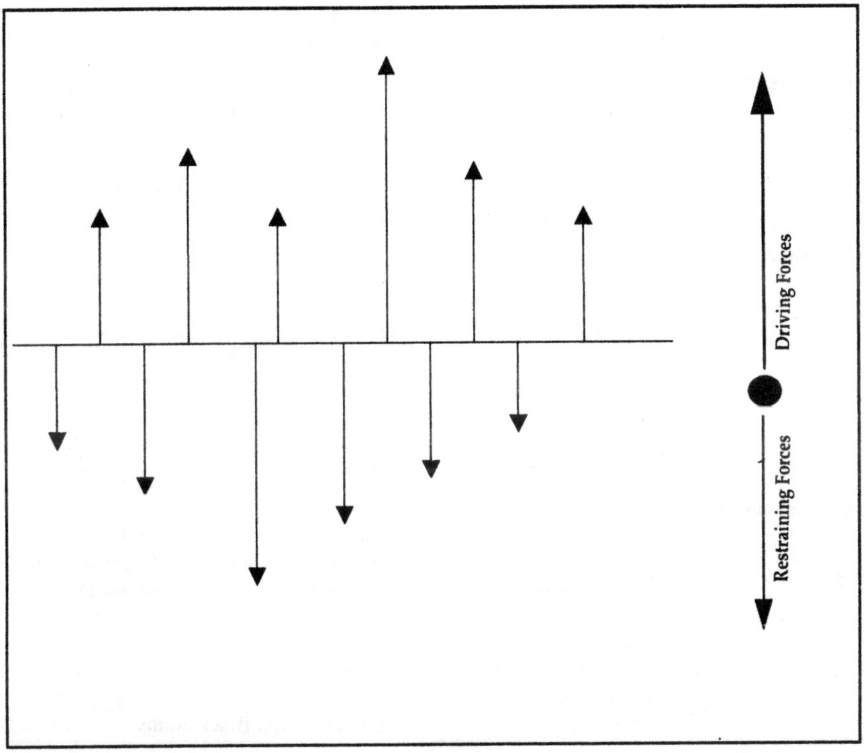

Source: Somesh Kumar, 1999.

Chart 1
Artisan Activities in Thiruvananthapuram District

Sl. No.	*Block*	*Grama Panchayat*	*Type of Artisan Activity Reported*
Parassala			
1.		Poovar	Handicrafts, wood works
2.		Thirupuram	Wood carving, handloom
3.		Chenkal	Wood carving, handloom
4.		Karodu	Wood work
5.		Kulathoor	Wood work
6.		Parassala	Handloom
Perumkadavila			
1.		Vellarada	Furniture works, wood curving
2.		Kunnathukal	Handicrafts [coir and cane]
3.		Kollayil	Handicrafts

Contd....

Sl. No.	Block	Grama Panchayat	Type of Artisan Activity Reported
4.		Perumkadavila	Handicrafts [coir and cane]
5.		Aryankodu	Wooden furniture, Bamboo, Palm leaf, Cane
6.		Ottasekharamangalam	Handicrafts, Cane, Bamboo
7.		Perumpazhuthoor	Furniture handloom
8.		Kallikadu	Wood mill
9.		Amboori	Wood mill
Nedumangad			
1.		Karakulam	Furniture
2.		Aruvikara	Wood, Handicrafts
3.		Vembayam	Gold works
4.		Anadu	Pottery, Handicrafts, Bamboo
5.		Panavoor	Cane furniture works
Kilmanoor			
1.		Pulimath	Carpentry, Goldsmith, Woodwork
2.		Karavaram	Furniture, Wood mill
3.		Nagarur	Furniture, Wood carving
4.		Kilimanoor	Carpentry
5.		Pazhya kunnummel	Carpentry
6.		Navayikulam	Carpentry
7.		Madavoor	Handicrafts [Bamboo, Coir, Palm leaf]
8.		Pallikal	Bronzesmithy, Stone carving
Chirayinkeezhu			
1.		Azhoor	Carpentry, Blacksmithy, Coir
2.		Chirayinkeezhu	Handicrafts, Cane, Coir, Woodworks
3.		Kadackavoor	Wood mills, Furniture, Goldsmithy, Carpentry
4.		Vakkom	Coir, Wood
5.		Anchuthengu	Coir
6.		Kizhuvilam	Coir
7.		Mudhakal	Carpentry and Blacksmithy
Trivandrum Rural			
1.		Vattiyoorkavu	Handicrafts, Coconut shell
2.		Kudappanakunnu	Furniture
3.		Ulloor	Wood mill, Coconut shell, Handicrafts
4.		Kadakampally	Carpentry, Coir mats, Handicrafts
Kazhakootam			
1.		Sreekariyam	Wood mill
2.		Pothencodu	Wood mill, Coir, Coconut shell
3.		Mangalapuram	Furniture, Carpentry, Handicrafts, Coconut shell
4.		Aandoorkonam	Furniture, Wood mills
5.		Kadinamkulam	Wood mills
6.		Kazhakootam	Wood industries, Coconut shell
Vellanadu			
1.		Kattakada	Furniture
2.		Vellanadu	Handicrafts [Bamboo]
3.		Poovachal	Wood mill, Furniture
4.		Aryanad	Handicrafts, Carpentry, Furniture

contd.

Sl. No.	*Block*	*Grama Panchayat*	*Type of Artisan Activity Reported*
5.		Vithura	Wood mill, Furniture, Coconut shell
6.		Kuttichal	Wood, Furniture
7.		Uzhamnalakal	Wood mill, Furniture
8.		Tholikode	Wood mills
Vamanapuram			
1.		Manikkal	Wood work, Carpentry, Wood printing
2.		Nellanad	Furniture
3.		Pullampara	Wood works, Furniture, Handicrafts
4.		Nandiyodu	Nothing
5.		Perighammala	Nothing
6.		Kallara	Wood mill, Plywood, Furniture, Handicrafts
7.		Pangode	Wood industry, Furniture
Nemom			
1.		Balaramapuram	Palm leaf, Carpentry, Handicrafts
2.		Pallichal	Furniture, Pottery, Handicrafts
3.		Malayinkeezhu	Bamboo, Furniture, Handicrafts
4.		Vilappil	Wood mill, Furniture, Handicrafts
5.		Vilavoorkal	Wood turning, Furniture, Handicrafts
6.		Kalliyoor	Handicrafts, Furniture
7.		Nemom	Wood works, Coconut shell, Handicrafts
Varkala			
1.		Vettoor	Wood mill, Coconut shell, Handicrafts
2.		Cherunniyoor	Wood mill
3.		Edava	Carpentry, Furniture, Wood mill
4.		Elakamon	Nothing
5.		Chemmaruthi	Furniture, Wood works, Cane, Coconut shell
6.		Manampoor	Wood mill
7.		Ottoor	Wood works
Trivandrum Corporation			
1.		Medical college	Wood works, Iron works, Furniture
2.		Kesavadasapuram	Handicrafts
3.		Vattiyoorkavu	Wood carving
4.		Vazhuthacaud	Wood, Fiber glass works
5.		Karamana	Wooden furniture
6.		Attukal	Handicrafts
7.		Pettah	Handicrafts [Sandal, Teak, Rosewood]
8.		Chakai	Handicrafts
9.		Perumthanni	Handicrafts
10.		Kuriathy	Wood mill, Furniture
11.		Sreevaraham	Carpentry
12.		Ambalathara	Handicrafts
13.		Kalippankulam	Handicrafts

Source: Development Report of Local Bodies.

Based on the above exercise it was decided to examine exhaustively the development projects of about 15-20 selected Grama Panchayats for 1997-98, 1998-99 and 1999-2000. It was noted from the consolidated report [prepared by the State Planning Board] on the Annual Plan 1997-98 of local bodies in

Thiruvananthapuram district that only very few projects were formulated for the economic development of marginalized groups like traditional artisans [Carpenters, Blacksmiths and Goldsmiths]. Since the Annual Plan 1999 - 2000 was in the implementation stage, financial and physical achievements could not be ascertained. Therefore it was decided to evaluate in detail only the artisan - related projects and their achievements for 1998 -99. Evaluation of few projects involving artisan activity / development like Handicrafts, Furniture, Handloom, Wood work, Coir, etc., indicated that projects were small in number as well as in terms of allocation / outlay; very few among them were successfully completed. Some of the projects proposed were not taken up, infact, for implementation. The results of the above evaluation exercise are summarized in Tables 1, 2, 3 and 4.

Table 1
Investment Pattern of Artisan-related Projects of Local Bodies [1997-98 and 1998-99] in Kerala

Type of Projects	*1997 - 98*			*1998 - 99*		
	No of Projects	*Grant in aid (%)*	*Total outlay (%)*	*No of Projects*	*Grant in aid (%)*	*Total outlay (%)*
Handicrafts	243	0.25	0.23	431	0.30	0.19
Handloom	46	0.22	0.21	140	0.25	0.22
Coir	111	0.17	0.30	202	0.33	0.36
Total	400	0.64	0.74	773	0.88	0.77
Productive Sector total	24932	33.93	42.32	32585	40.66	38.34

Source: Government of Kerala, State Planning Board, Economic Review 1999.

Table 2
Grant in aid to Local Bodies for Financing Artisan Projects (1997-98) Thiruvananthapuram District

Projects	*Sectoral Distribution (%)*		
	General	*Special Component*	*Women component*
Furniture	0.05	0.00	0.00
Handicrafts	0.62	0.40	3.37
Handloom	1.59	4.99	0.59
Coir	0.15	0.09	1.27
Total	2.41 (34.43)	5.48 (23.22)	5.23 (68.05)

Source: Annual Plan (1997-98) Local Bodies.
Figures in parenthesis indicate the percentage share of grant-in-aid distributed for productive sector development projects.

Investment pattern of artisian related projects of local bodies in Kerala for the year 1997-98 and 1998-99 is summarized in Table 1. Though the number of productive sector artisan projects has increased from 400 to 773, their total outlay in percentage terms has remained stagnant; grant-in-aid to local bodies for artisan projects has increased from 0.64% to 0.88%. The trend was more or less the same for productive sector as a whole. The sectoral distribution of grant-in-aid to local bodies in Thiruvananthapuram district for 1997-98 is given in Table 2. The predominance of handloom projects in the general and special component sectors is obvious. Taken together, the artisan projects (furniture, handicrafts, handloom and coir) accounted for 2.41 per cent of the general productive sector share (34.43%) in grant-in-aid distributed to local bodies. Under the special component plan the relative share of grant-in-aid for artisan projects (5.48%) was higher than its relative share (2.41%) in the general sector and also in relation to the total share of grant-in-aid distributed to the sector (23.22%) as a whole.

An evaluation of artisan projects in 23 Grama Panchayats for the year 1998-99 indicate that out of 28 artisan projects undertaken by them only 14 have been completed within the year; 12 projects were not completed and 2 were not implemented at all as may be seen from Table 3. Comparing Tables 3 and 4 it is understood that one project not implemented in 1998-99 was taken up during the year 1999-2000. There are no specific references in the plan proposals for the year 1999-2000 with regard to the project not completed in 1998-99. The popular artisan projects in many Panchayats over the years have been furniture, wood turning, handloom and handicrafts. Eventhough handicrafts consist of a number of small crafts making use of a variety of raw materials involving different levels of skill, such details are not available from the plan documents. How far these projects have been helpful to traditional artisans is also not known, given the high degree of occupational inmobility persisting in the artisan labour market.

According to a survey conducted in 1992 there were 17669 rural artisans in Thiruvananthapuram district, which is the highest that comes to 18.85 per cent of the total rural artisan population (93733) in Kerala. It is doubtful whether the issues of those rural artisans in the district are adequately addressed under the people's plan.

From the above review it may be noted that it would be desirable to formulate specific projects for the development of traditional artisans like carpenters, blacksmiths and goldsmiths in selected clusters. Organized activity of traditional artisans could be promoted under local level planning more effectively and in a participatory manner through the cluster approach and as part of an artisan component or plan like SCP/TSP. Since traditional artisans are not covered so far by a group approach as in the case of scheduled castes and scheduled tribes, the cluster approach seems inevitable. The cluster approach could be taken as a forerunner or a pilot programme towards an artisan component in multi-level planning. It was against the above background that we have proposed a model artisan plan based on the participatory and cluster approaches.

POTTERY UNIT: AN EVALUATION

Against the above background, we decided to undertake an ex-post evaluation of an artisan project through field visit. The project selected at random for evaluation is a pottery unit (Project No. 21 in Table 3) supported by Vilavoorkal Grama Panchayat, Nemon Block, in 1998-99. The project, proposed by the Task Force on Industries (not by the Grama sabha) was approved on 27-10-1998 by the subject

Table 3

Particulars of Artisan Projects in Selected Grama Panchayats (1998-99) Thiruvananthapuram District (Amount in Rs.)

Sl. No.	Name of Panchayat	Name of project	Grant-in-aid	Own fund	State Sponsored Scheme	Centrally sponsored Scheme	Loan from Coope-ratives	Loan from financial institutions	Voluntary Service/ contri-bution	Share of Bene-ficiary	Others	Total outlay	Funds allocated (P) Financial achievement as on 31 March (e)	Project Completed (Y) Not completed (N)
(1)	(2)	(3)	(4)	(5)	(6)	(7)	(8)	(9)	(10)	(11)	(12)	(13)	(14)	(15)
1.	Kazhakuttom	Furniture Mart				14,000		76,000				90,000	P	Y
						14,000		76,000				90,000	E	
2.	Vellarada	Wood Turning Industry	1,10,375											N*
3.	Kalliyoor	Handicrafts	50,000							14,000		64,000	P	N
			30,000									30,000	E	
4.	Vithura	Handicrafts Unit	10,000							20,000		30,000	P	Y
			10,000							20,000		30,000	E	
5.	Kallikkadu	Handloom Production & Training	62,500						15,000			75,000	P	N
			-									-	E	
6.	Kollayil	Women's Handloom Weaving	1,00,000							90,714		1,90,714	P	N
			99,376									99,376	E	
7.	Perumpazhathoor	"	1,19,000							51,000		1,70,000	P	N
			35,000							15,000		50,000	E	
8.	"	"	15,000-							5,000-		20,000-	PE	N
9.	Vellarada	Handloom	3,63,217							3,66,786		7,30,003	P	N
			960							-		960	E	
10.	Vizhinjam	Handloom	1,00,000							1,00,000		2,00,000	P	Y
			1,00,000							1,00,000		2,00,000	E	
11.	Athiyannoor	Handloom Training	1,00,000									1,00,000	P	Y
			70,000									70,000	E	
12.	Balaramapuram	Handloom	2,34,030							56,550		2,90,580	P	Y
			1,04,910							25,350		1,30,260	E	
13.	Kazhakpptam	Weaving Unit	-			38,000		2,82,000				3,20,000	P	Y
						38,000		2,00,000				2,38,000	E	

contd.

(1)	(2)	(3)	(4)	(5)	(6)	(7)	(8)	(9)	(10)	(11)	(12)	(13)	(14)	(15)
14.	Anad	Weaving	35,000							50,000		85,000	P	N
		production &	-							-		-	E	
		training												
15.	Kizhuvilam	Handloom	4,56,000	5,020						2,30,975		6,91,995	P	Y
		Jyothi	4,56,000							-		4,56,000	E	
16.	Parassala	Pottery	55,000							55,000		1,10,000	P	Y
			55,000							-		55,000	E	
17.	"	Palm Industry	1,00,000							1,00,000		2,00,000	P	Y
			1,00,000							-		1,00,000	E	
18.	Kallikadu	Mat making	16,500						16,500			33,000	P	N
			12,100						12,100			24,200	E	
19.	"	"	30,000						10,000			40,000	P	N
			26,000						7,580			33,580	E	
20.	Malayinkil	Cane Industry	1,85,700							62,500	25,000	273200		N*
			-											
21.	Vilavoorkal	Pottery Unit	15,000						21,750			36,750	P	Y
			15,000									15,000	E	
22.	Anad	Cane Industry	80,000							1,21,800		2,01,800	P	N
			47,000							71,550		1,18,500	E	
23.	"	"	20,000							27,250		47,250	P	N
			5,000							6,810		11,810	E	
24.	Vettor	Bamboo Unit	8,940							1060		10,000	P	
			3,576							424		4,000	E	N
25.	Edava	Coir Products	50,000							50,000		1,00,000	P	Y
			39,500									39,500	E	
26.	"	Women's Co-op Society	15,000							9,500		24,500	P	Y
			15,000							9,500		24,500	E	
27.	Ottoor	Coir Co-op Society	1,00,000									1,00,000	P	Y
			1,00,000									1,00,000	E	
28.	Kadackavoor	Coir Industry	50,000									50,000	P	Y
			48,000									48,000	E	

Source: Statement of Projects (1998-99) of Grama Panchayats.
*Not implemented

Table 4

Particulars of Artisan Projects in Selected Grama Panchayats (1999-2000) Thiruvanathapuram District **(Amount in Rs.)**

Sl. No.	*Name of Panchayat*	*Name of Project*	*Grant-in-aid*	*Panchayat Expenditure*	*Direct Expenditure*	*Total*	*No. of beneficiaries*	*No. of working days*
(1)	(2)	(3)	(4)	(5)	(6)	(7)	(8)	(9)
1.	Vizhinjam	Furniture unit	75,000		25,000	1,00,000	5	
2.	Vellarada	Wood Turning industry	1,10,375 (98-99)		1,30,125	2,40,500	30	
3.	Kilimanoor	Bamboo	25,000		27,240	52,240	5	
4.	Perumpazhathoor	Pottery	70,000		28,000	98,000	28	730
5.	Parassala	Palm leaf	1,00,000		25,000	1,25,000	5	920
6.	”	Clay	50,000		50,000	1,00,000		
7.	Chemmaruthi	Handicrafts	50,000	22,000		72,000		
8.	Vellarada	Handloom	3,62,351 (98-99)		3,66,786	7,29,137	30	
9.	”	”	2,00,115 (98-99)		2,07,392	4,07,507	16	
10.	”	”	8,06,100		8,06,100	16,12,200		
11.	Balaramapuram	Handloom	70,000		70,280	1,40,280	14	
12.	Pallichal	Handloom		50,000		50,000		

Source: Statement of Projects (1999-2000) of Grama Panchayats.

committee (BLEC) for Small Scale Industries, consisting of Senior Manager, SIDCO, MD, Bee fed and Assistant District Industries Officer, Neyyattinkara with a total outlay of Rs. 36750.

Clusters of traditional village potters facing problems of unemployment and poverty spot Vembadatheri, in Choozhattukotta Ward of Vilavoorkal Panchayat. The objectives of the pottery unit were:

1. Rehabilitation of pottery as a village craft
2. Facilitation of new generation in to pottery
3. Removal of poverty among traditional potters
4. Generation of sustained employment
5. Enhancement of productive capacity, and
6. Provision of marketing facility to artisans in pottery.

Beneficiaries of the unit-10 families of traditional village potters-inclusive of women, were identified from among families below the poverty line, through 'Ayalkoottom'. Potters in the family are traditional/skilled artisans carrying on the production of utensils, flower-pots and a variety of art goods in clay. A chairman and convener were proposed by the Ayalkoottom to initiate the project through infrastructure renovation of a 15-year-old pottery unit, under the organizational framework DWCRA. Under the supervision of the project leaders-chairman and convener-other activities were coordinated viz, registration of the unit, agreement with the owner of the land (where workshed, kiln and drying yard are located), development of infrastructure, training and awareness, procurement of raw materials, allocation of work among skilled labour (men and women), quality control and management of marketing through co operatives, public markets, exhibition, door delivery etc. As per the calendar of activities proposed, barely five months were visualized for the whole process. Monitoring of the unit was left with a body consisting of elected members (project leaders) of the unit, panchayat official (VEO, being the implementing officer) and members of the task forces.

The economics of the unit as per the proposal is summarized in Tables 5 and 6.

Table 5
Village Pottery Unit: Working Capital (turnover period one month)

Raw materials	*Plan funds (Rs.)*	*Beneficiary share (Rs)*
Clay	5500	-
Sandy soil	1200	-
Mud soil	7000	-
Other materials	1000	-
Rent	-	5000
Renovation of Shed, Kiln	-	5000
Tools	-	3000
Labour (transportation)	-	2000
Project implementation and monitoring	300	-
Total	15,000	15,000
Total outlay	30,000	

Source: Project Proposals, Vilavoorkal Grama Panchayat 1998-99.

Table 6
Pottery Unit: Cost, Revenue and Profit (per month)

	Particulars	*Amount (Rs.)*
1.	**Cost**	
	Raw materials	14700
	Implementation & monitoring	300
	Commission for marketing	4800
	Wages	22500
	Total	42,300
2.	**Revenue**	
	Pot 1000 x 14	14,000
	Flower pot 1000 x 10	10,000
	Pot (small) 2000 x 7	14,000
	Pan 2000 x 5	10,000
	Total	48,000
3.	**Profit (2-1)**	5,700
4.	**Income per head (10 workers)**	570

Source: Project Proposals, Vilavoorkal Grama Panchayat, 1998-99.

EX-POST EVALUATION OF POTTERY UNIT: SOME OBSERVATIONS

Pottery is as ancient as human civilization and is perhaps the first craft which rose from its utilitarian origins to become a thing of beauty and artistic expression. It is among the important village industries in India and there is hardly a village that does not have its own potter. About 15 lakhs potters are engaged in the manufacture of pottery in the unorganized sector. Their livelihood depends on pottery as their main occupation [NISTADS, 1996]. The potters employ traditional skills for making a large variety of earthenwares required for domestic use and other purposes in and outside the village. The availability of suitable clay, highly skilled potters, finance, tools and technology and ready market for the wares are essential requirements for the healthy development of this craft. Pottery industries can be categorized into three groups viz, large/medium-sized industries, small-scale industries and individual-owned units/enterprises. Small and individual- owned units face problems in upgradation of techniques/technology, acquisition of tools, getting raw materials, finance, marketing etc.

A recent study [Solanki and Qureshi, 1998] has identified in Kerala 74,750 potters engaged in red clay works in which only 9.90% belong to the non-traditional category, indicating that the craft continues to be traditional and family-based. The plight of traditional potters in the state, as in other states, leaves much to be desired with majority of them struggling beneath the poverty line and under the clutches of middlemen for want of effective/sustained state intervention. Though the market for traditional pottery items has been dwindling with the advent of metallic/plastic manufactures, the market for non-traditional and utility items like flowerpot/vase has been burgeoning. However, for want of elastic/efficient production base and institutional support, majority of the potters continue in the poverty stricken brackets of society. Many of the development interventions have by- passed traditional potters scattered in village panchayats and suburbs of metropolis. It is hoped that our observations on the pottery unit, being evaluated, would also generate useful points for action towards the rehabilitation of traditional artisans in the state.

The views of the beneficiaries and our observations on the conceptualization, implementation and monitoring of the project are summarized below.

1. Though majority of the potters have attended Grama sabha, their participation has not been "informed", as the idea of the pottery unit was not floated by them but by the task force.
2. Mainly the external experts conceived the pottery unit, with little attention to the social environment under which the traditional potters operate. Following the renovation of a family temple [in the courtyard of which the unit has been located], the pottery unit had to be shifted, within few months to be relocated in the narrow household premises of the project leader.
3. As the production process has come to a halt, project beneficiaries (especially men) have engaged themselves in alternative wage employment outside the system, leaving the task of reestablishing the unit and resuming production to the shoulders of the project leaders. Organizational arrangement of the unit has failed to keep all the beneficiaries equally responsible for the sustenance of the system. It is to be noted that the renovation process in the temple, involving the shifting of the pottery unit, was executed without any difference of opinion; in fact it was done under a shared common religious belief, upholding its priority over economic considerations.
4. The organizational pattern of the unit leaves little scope for seeking credit support from institutional sources to reestablish it in the new premises. The emerging environment is more likely to take the potters closer to village money lenders/intermediaries in the marketing channels.
5. Supply irregularities led to loss of regular customers and fair price for products. As women employment is maximized when production units are closer to the household, the shifting of the unit and discontinuities in production leave majority of them unemployed.
6. A pottery production cooperative or trust (as in the case of 'The Potters Shed' near Kodaikanal, Tamil Nadu) would have secured continuity to the pottery unit and access to alternative sources of credit/marketing.

To sum up, the initiative of the Grama Panchayat towards rehabilitation of traditional potters has not been adequate to render them with more employment, income and enhanced production/productivity. The small size of grants-in-aid by itself was not the only limitation; the inability to mobilize supplementary finance, dependency of the production system on the social environment of the group and lack of a unifying force to hold the beneficiaries together, have been neither foreseen nor guarded against adequately, in the case of the village pottery unit under evaluation. The DWCRA framework is grossly inadequate for the process; an organizational alternative capable of sustaining the production system has to be thought of, given the production infrastructure already available, assured market for products and supply of raw materials.

TOWARDS FORMULATING AN ARTISAN PLAN: DEVELOPMENT SEMINAR AND PRA WORKSHOP: A DISCUSSION

As a logical sequence of the review of development plans of selected local bodies (mainly Grama Panchayaths) in Thiruvananthapuram district, we decided to organize a development seminar and a PRA workshop to sensitize artisans towards the idea of an 'artisan plan'. The intensive scanning of the Development Reports of selected grama Panchayats with artisan clusters indicated that Soolamkudy, Ward IX of Poovar Grama Panchayat in Parassala Block would be a suitable area for PRA. Preliminary enquires and discussions with communal organizations of artisans also facilitated the choice of the study area. A socio-economic profile of the study area is given below.

SOCIO-ECONOMIC PROFILE OF POOVAR GRAMA PANCHAYAT AND SOOLAMKUDY

Thiruvananthapuram, the Southern most district and the administrative capital of Kerala State, has 84 Grama Panchayats, 12 Block Panchayats, 4 Municipalities, and one Municipal Corporation, besides the District Panchayat. Poovar Grama Panchayat is located in Parassala Block. Geographically the Panchayat is surrounded by Karumkulam Panchayat in the west, Neyyar in the east, Thirupuram Panchayat in the North and Arabian Sea in the South. Poovar Grama Panchayat, which came into existence in 1969, was carved out of Thirupuram Panchyat. According to the Development Report (1997-2001), its total geographical area is 7.34 sq. k.m. and its population is 19777 (1991-Census). Poovar Panchayat is divided into 9 wards. Literacy rate of the Panchayat (86%) is lower than that of the district (89%). It has typical characteristics of a rural setting in Kerala. The village population is mainly engaged in agriculture and fishing for their livelihood. The population comprises of farmers, fishermen, scheduled caste and other backward communities. Majority of the people in the Panchayat are either cultivators or agricultural labour. Construction, fishing etc. constitute other activities of the people. The work-participation rate is 48% for male and 12.3% for female population. Particulars of major economic activities, work participation rate, infrastructure facilities like housing, electrification, drinking water, schools, medical institutions, public distribution system etc as well as the type of social and cultural institutions in the Panchayat area are summarized in Tables 7 to 10.

Soolamkudy, the study area chosen is ward IX of Poovar grama panchayat. Traditional artisan households of carpenters, blacksmiths and goldsmiths are found well spread out all over the ward. Most of the traditional artisans are following hereditary occupations like carpentry, blacksmithy, and goldsmithy. Since they possess very little agricultural land, formal education, income-yielding assets etc., they continue as traditional workers to earn a livelihood. Hence there is scope for

Table 7
Sex-wise Distribution of Population: Poovar Grama Panchayat, 1991

Population	*Male*	*Female*	*Total*
Schedule Caste	1149	1169	2318
Schedule Tribe	16	19	35
Others	8215	8187	16402
Total	9380	9375	18755
Literacy rate	88.95	83.08	86.00

Source: Panchayat Level Statistics, Thiruvananthapuram District, Department of Economics and Statistics, Government of Kerala, 1996.

Table 8
Activity-Status of Working Population: Poovar Grama Panchayat, 1991

	Activity status	*Male*	*Female*	*Total*
1.	Cultivators	802	37	839
2.	Agricultural workers	1647	173	1820
3.	LFP	45	18	63
4.	Mining and Quarrying	1	0	1
5.	M.P.S.H	73	33	106
6.	M.P.S.O	140	41	181
7.	Construction	108	8	116
8.	Trade And Commerce	218	46	264
9.	T.S.C.	187	9	196
10.	Other Services	736	429	1165
	Total	3957	794	4751

Source: Panchayat Level Statistics, Thiruvananthapuram District, Department of Economics and Statistics, Government of Kerala, 1996.

Table 9
Work Participation Rate: Poovar Grama Panchayat, 1991

Types of workers	*Male*	*Female*	*Total*
Main workers	4335	820	5155
Marginal workers	167	340	507
Total workers	4502	1160	5662
Work Participation Rate	48.00	12.37	30.19

Source: Panchayat Level Statistics, Thiruvananthapuram District, Dept. of Economics and Statistics, Government of Kerala, 1996.

developing a model artisan plan with the active participation of traditional artisans inhabiting the ward.

DEVELOPMENT SEMINAR AND PRA WORKSHOP: A DISCUSSION

As already stated, a development seminar and PRA workshop were organized in Soolamkudy, on Sunday 13th February 2000. The venue of the assembly was the courtyard of a local temple in Soolamkudy. Sixty eight traditional artisans

Table 10

Housing and other Infrastructure: Poovar Grama Panchayat, 1991

Infrastructure	*Characteristics*	*No*
1. Housing	Concrete	1024
	Tiled	1010
	Thatched	2788
	Total	4822
2. Electrification	Electrified	1948
	Non-electrified	2874
	Total	**4822**
3. Public Distribution System	Ration Shops	14
	Card Holders	3250
	L.P.G	36
	L.P.G Agency	1
	Total	**3301**
4. Schools	Govt. L.P. School	3
	Govt. High School	1
	Private High School	1
	Total	**5**
5. Medical Institutions	*(a) **Allopathy***	
	Government Hospital	0
	Private Hospital	1
	Community Health Centre	1
	Family Welfare Centre	10
	*(b) **Ayurveda***	
	Private Hospital	1
	Private Dispensary	1
	*(c) **Homeopathy***	
	Private Hospital	1
	Government dispensary	1
	Private Dispensary	1
	Total	**17**
6. Scheduled Caste/Tribe Colonies	SC colony	2
	ST colony	0
7. Drinking Water	Public Tube well	5
	Public Wells	12
	Public Tank	2
	Public tap	18
	Total	**37**
8. Social and Cultural Institutions	Public Library	2
	Anganavady	3
	Balavady	3
	Adult Education Centre	2
	Community Hall	2
	Sports clubs	8
	Arts clubs	4
	Community Radio sets	2
	Community TV sets	2
	Cinema Theatre	1
	Total	**29**

Source: Panchayat Level Statistics, Thiruvananthapuram District, Dept of Economics and Statistics, Govt. of Kerala, 1996.

participated in the seminar/workshop which was inaugurated by the President of the gramapanchayat. In order to collect basic socio-economic information about the participating artisans, a structured questionnaire was circulated in the seminar session. In the seminar session there were meaningful discussions and interaction on the content, objectives, methodology and the process involved in participatory planning at the panchayat level. There was also sharing of experiences in grama sabha, task forces and panchayat level development seminar. The advantages as well as the limitations of the People's Plan were discussed at length. Majority of the participants shared the view that the existing projects in the panchayat do not address effectively the development problems of traditional artisans. The need for a new approach to the issues of traditional artisans was emphasized. The above view facilitated our exercise towards developing an artisan plan (in the context of local level planning), which is exclusively meant for the benefit of traditional artisans.

The workshop session was used to develop an artisan plan through Force Field Analysis in Participatory Rural Appraisal (PRA). Force Field Analysis (FFA) is a technique developed by Kurt Lewin (1951) to visually identify and analyse forces affecting a problem situation so as to plan a positive change. It has been used in diverse fields ranging from organizational change to self-development. Its visual character, simplicity, suitability for group work and applicability in planning for change makes it a potential tool with wide application in PRA. According to FFA any situation or performance can be viewed as a state of temporary equilibrium. It is caused by two sets of opposing forces, one set comprises of forces that try to bring change referred to as driving, facilitating or positive forces. The other set of forces, which try to maintain the status quo is termed restraining, resisting or negative forces. FFA involves the following steps, as illustrated in Flipchart: Force Field Analysis (Chart 2).

The process/steps involved in FFA was explained to the participants so as to initiate them into the workshop. Participants were able to prepare seasonal employment calendars (Chart 3), map the dwellings of artisans, identify driving and restraining forces (as illustrated in (Fig. 2) and react meaningfully to local problems. The driving forces as identified by the participants are infrastructure facilities, meetings of the grama sabha, awareness about development issues, credit facilities, voluntary services, opportunities for forming co-operatives and Panchayath level development plan. They have identified the restraining forces to be competition from substitutes, shortage of raw material, lack of education, changes in technology, inflow of labour from neighboring states, contract system and shortage of capital. Both types of forces were also ranked in the order of priority/ strength with a view to further strengthen the most powerful forces, according to the foremost objective of the assembly, namely the formulation of an artisan plan.

The idea of an artisan cooperative for the production of wood/iron based building materials (which is capable of giving regular and gainful employment to them), was floated by senior artisans in the assembly/workshop. According to the artisans, the strongest driving force for a change in the current situation is the

Chart 2: Force Field Analysis

Step I: Set out on a flipchart the following:

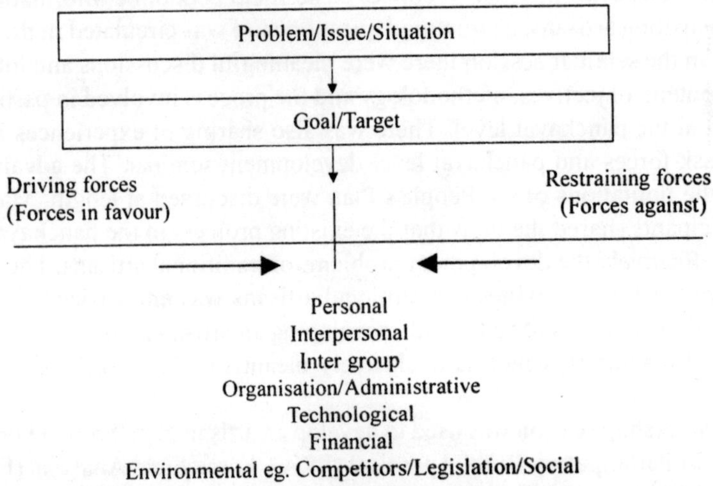

Step II: Fill in the problem/issue/situation now and goal/ target boxes.

Step III: Make a list of all the forces that might restrain or oppose a solution/change. Rank them in order of the forces that are the strongest and therefore might be most difficult to overcome. List them on the righthand side of the chart. It is often helpful to indicate the strength of the restraining force by drawing an arrow under each force, the length representing the degree of opposition that might be encountered.

Step IV: Then make a list of the driving force-those that encourage and facilitate the change. Rank them and depict them, as in step III, on the left hand side of the chart. The aim is to capitalize on those that will make the greatest contribution to resolving the problem/ bringing about the decided change.

Source: Frances and Roland, 1998.

Chart 3

Seasonal Employment Calendar of Artisans, Soolamkudy-1999, Thiruvananthapuram District

Month	*Carpenters*	*Blacksmiths*	*Goldsmiths*
January	18 days	7 days	10 days
February	17 days	9 days	12 days
March	9 days	8 days	8 days
April	12 days	5 days	8 days
May	14 days	10 days	7 days
June	11 days	9 days	10 days
July	12 days	7 days	12 days
August	15 days	5 days	14 days
September	18 days	7 days	15 days
October	15 days	5 days	8 days
November	16 days	9 days	13 days
December	12 days	5 days	12 days
Total	169 days	86 days	129 days

Prepared by:	S. Sree Kumar Panikkanvilakam Soolamkudy	J.S. Suresh Kumar Melethilputhan veedu, Soolamkudy	V. SunilKumar ThekathilVeedu Soolamkudy

Figure 2: Formulation of Artisan Plan: Force Field Analysis

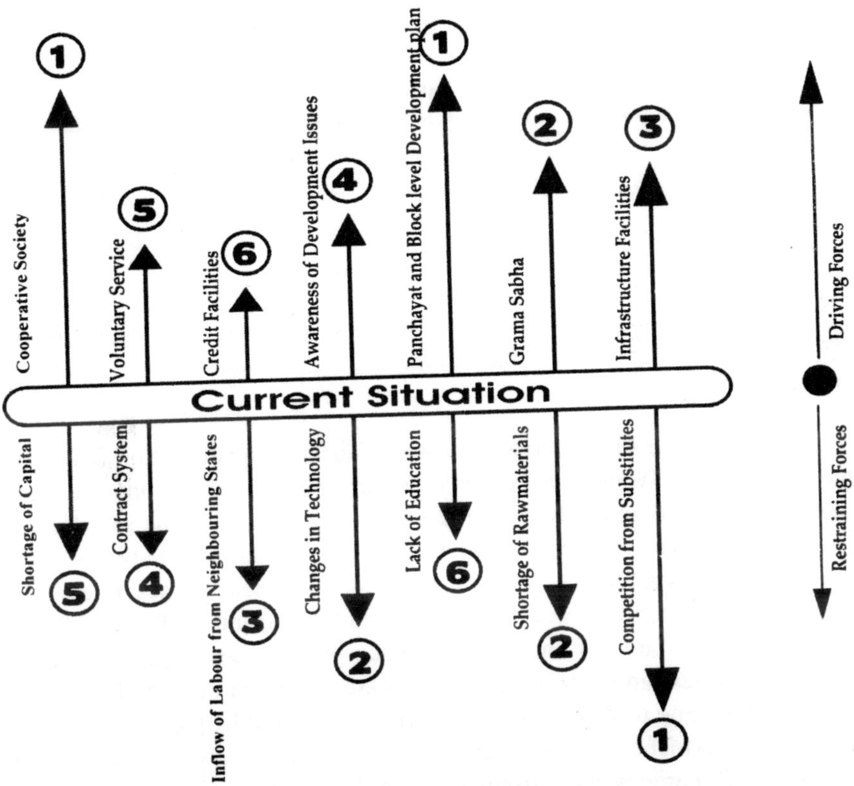

The Encircled Numbers Represent Significance Rank of Factors

Participants	:	**Rural Artisans (Carpenters, Blacksmiths & Goldsmiths)**
Grama Panchayat	:	**Poovar, Thiruvananthapuram, Kerala**
Place	:	**Mutharammankovil, Soolamkudy**
Date & Time	:	**13th February, 2000, 4 P.M.**

avenue to organize cooperative form of enterprises under the participatory people's plan of the grama panchayat. Participants of the workshop held the view that such a cooperative would have adequate business under the new dispensation of Panchayati-Raj in which a number of activities and institutions have been transferred to local bodies. A system of priorities or even reservation of work orders, marketing etc could be thought of for the proposed cooperative. The organization of cooperatives has become quite common in local planning and economic development and instrumental in translating the objectives of local-level development in a participatory framework.

SOCIO-ECONOMIC STATUS OF RURAL ARTISANS OF SOOLAMKUDY

Educational status of rural artisans in the study area in given in Table 11.

Table 11
Educational Status of Rural Artisans in Soolamkudy, Poovar Grama Panchayat

Artisans	*Educational Status*						
	Illiterate	*Below 5th Std*	*Std 6-10*	*PDC*	*Degree*	*Technical*	*Total*
Carpenters	2	14	45	3	0	0	64
Blacksmiths	0	1	0	0	0	0	1
Goldsmiths	0	0	2	0	1	0	3
Total	2	15	47	3	1	0	68

Source: Primary Survey.

Majority of the Artisans (68%) possess education qualification between 6th and 10th standard. Twenty two percent of the artisans have educational background below 5th standard and only two artisans are illiterate according to information available from primary data. It may therefore be concluded that rural artisans generally possess secondary level of education and their literacy rate (97%) is higher than the general literacy rate of the district (89%) and the Panchayat (86%). Hereditary background of employment, place of work, average annual days of active employment and average daily wages are examined in Table 12.

It is interesting to know that all the artisans covered by the study have taken up their employment as a continuation of traditional family employment. Only very few (10%) among the artisans perform their work in their own household. Majority of the artisans (73.53%) perform their work outside their households work sheds mainly as daily wage earners indicating lack of enterprise and initiative for self employment. It was also observed that few artisans (16.18%) work under local contractors on a regular basis. About 47% of the artisans normally work within the Panchayat and others in the neighboring Panchayats. We have observed wide variations in the average annual working days reported by the artisans. Average number of working days per year generally varies between 100 and 200 days in the opinion of an absolute majority (80.88%). About 42.65% have stated average annual working days to be ranging between 150 and 200. Average daily wages of most of the artisans (69.12%) were found to be ranging between Rs 100-150 which is an indication that the level of wages do not reflect adequately or correspond to the skill of artisans. Generally we expect higher wages for higher skill. From the data relating to the average daily wages of rural artisans in the study area and also secondary information on wage structure of rural workers in the district (relating to skilled carpenters and masons on the one hand and agriculture labour and construction workers, considered to be unskilled on the other hand), we have observed that in the determination of rural wages, the level of skill is unimportant. Our findings endorse fully the interpretation of the results of the Report of Wage

Table 12

Employment and Wage level of Rural Artisans in Soolamkudy

Artisans	*Type of Artisans*		*Place of Employment*			*Within Local*	*No. of Working Days Per Year*				*Daily Wages (Rs)*		
	T	*NT*	*H*	*O*	*C*	*Panchayath*	*<100*	*100-150*	*150-200*	*>200*	*<100*	*100-150*	*>150*
Carpenters	64	0	6	48	10	29	2	25	28	9	5	47	12
Blacksmiths	1	0	1	0	0	0	1	0	0	0	1	0	0
Goldsmiths	3	0	0	2	1	3	0	1	1	1	2	0	1
Total	68	0	7	50	11	32	3	26	29	10	8	47	13

Source: Primary Survey.

T: Traditional NT: Non- traditional

H: Household 0 : Outside C : Under contractor

Structure published by the Government of Kerala (Report of Wage Structure, 1982-1992, Department of Economics and Statistics, Government of Kerala 1995).

Table 13
Subsidiary Employment, Mobility and Tools of Rural Artisans in Soolamkudy, Poovar Grama Panchayat

Artisans	*Having subsidiary employment (Nos)*	*Doing overtime work (Nos)*	*Interested in Gulf employment (Nos)*	*Those occasionally hiring tools*	*Finance for Purchase of Tools*			
					Loan	*Subsidy*	*Grant*	*Own Source*
Carpenters	8	3	42	13	5	0	0	59
Blacksmiths	0	0	1	0	0	0	0	1
Goldsmiths	0	0	2	1	0	0	0	3
Total	8	3	45	14	5	0	0	63

Source: Primary Survey.

Though there exists subsidiary employment in the case of a large section of the rural work force, artisans are an exception; normally they stick to their main employment and seldom undertake overtime or secondary work for additional income. None of the participating artisans have benefited from the scheme for the Supply of Improved Toolkits to Rural Artisans (SITRA), offered by the Union Government (See Table 13). We have analyzed the consequences of the inflow of outside artisans to the locality in terms of the trends in their wages as reported by the local artisans. The opinion is found to be sharply divided on the levels of wages of migrants vis-à-vis the natives (See Table 14).

Table 14
Opinion of Rural Artisans of Soolamkudy on Migrant Labour, Occupational Mobility and People's Plan

Opinion	*Artisans*			
	Carpenters	*Blacksmiths*	*Goldsmiths*	*Total*
Number of artisans stating labour immigration	45	1	2	48
Number of artisans stating wage of immigrant labour to be				
High	25	0	2	27
Low	20	1	1	22
Same	3	0	0	3
Compared to local wages.				
Number of artisans stating the possibility of horizontal occupational mobility from				
(a) carpentry to blacksmithy	34	0	·1	35
(b) blacksmithy to carpentry with some initial training	35	1	0	36
No of artisans stating				
• little knowledge/participation in People's Plan, and	49	1	3	53
• the need for an Artisan Plan similar to SCP/TSP	55	1	3	59

Source: Primary Survey, Development Seminar and PRA Workshop.

About 42 per cent of the reporting artisans have stated lower wages (compared to their own) of the migrants. Conversely a higher percentage (52) of the native artisans consider that the wages of the migrants are higher than that of their own. Since the opinion is, as stated above, sharply divided, no meaningful conclusion seems emerging. Further in -depth study alone can unravel the process involved. Since the local wages are not very much linked to skill and also relatively low, most of the artisans are willing to take up employment in Gulf countries where they expect higher wages. We have posed the question whether it is possible, given preliminary training, to move from carpentry to blacksmithy and vice versa. Majority of the artisans (52%) were of the opinion that it is possible to move along the above path given some training.

It is generally observed that members of the marginalized communities are reluctant (lack initiative) to enroll themselves in service institutions, political parties, cooperative society etc. In the above background, the opinion of rural artisans was sought on the question. Surprisingly many rural artisans do not have membership in government-sponsored schemes like Artisans Benefit Scheme or artisans and other skilled workers welfare fund scheme. Their membership in communal organizations, insurance schemes, trade unions and political parties is also nominal (See Table 15).

Table 15
Membership of Rural Artisans of Soolamkudy in Service Institutions and Participation in Peoples Organisations

Artisans	*Membership/Participation in*						
	*ABS**	*Communal Organisations*	*Life/other Insurance*	*Trade Union*	*Cooperative Society*	*Political party*	*Grama Sabha*
Carpenters	3	5	10	8	41	11	46
Blacksmiths	0	0	0	0	0	0	0
Goldsmiths	0	1	1	0	2	0	2
Total	3	6	11	8	43	11	48

Source: Primary Survey.
*Artisans Benefit Scheme, Government of Kerala.

The only exception is membership in cooperative society and participation in grama sahba, where 63 per cent and 70.58 per cent hold membership/attend the meeting, respectively. The reported situation of lack of enrollment in government schemes and popular organisations calls for regular awareness programs and facilitation strategies for the promotion of the socio-economic interest of rural artisans.

CONCLUSIONS AND SUGGESTIONS

Rural artisans comprising carpenters, goldsmiths, blacksmiths bronze/coppersmiths, masons and potters are one of the most depressed classes in the society. Traditional skill is their only asset. With the advancement of science and technology, their trades characterized by traditional skill and primitive tools are being ousted by the factory sector. The advent of the new building technology, with its shift from wood to concrete has considerably reduced the employment opportunities of the carpenters. Mechanisation and modernization have made deep inroads into the field of goldsmithy, blacksmithy, bronzesmithy, pottery and shoe making. Though highly skilled workers, they lack entrepreneurial talents, leadership qualities and capital. Nearly 50% of them work as wage earners. The rest are household enterprise workers or manage own account non- household single worker establishments. Whether self- employed or wage earners, the artisans are heavily exploited. Being economically and socially weaklings and numerically few they have little bargaining power and all the development efforts have largely bypassed them (Committee for Artisan Development, 1979). Considering the peculiar problems of artisans, the Committee suggested a package of services through the Kerala Artisans Development Corporation. The Corporation envisaged a ten year development programme consisting of 3200 primary production units, 40 development centers, supply stores etc. However, it could promote through its development centers only 6 primary production units in Kottayam and 3 units in Kannur. A close look at the Corporation reveals that its primary objective (organization of production units) remains totally neglected and that state support (budget allocations) is also receding. No wonder, the Corporation has failed to make any impact on artisan issues. State initiative towards improvement of artisans of Kerala (like the artisan benefit scheme, Handicrafts Development Corporation, the Scheme for the Supply of Improved Toolkits to Rural Artisans etc has also failed to address the issues effectively. Though Kerala is much ahead of others in terms of social security and welfare enactments, the benefits of such progressive measures have failed to percolate to the artisan groups in any significant measure (Sankaran 1999).

People's Campaign for Ninth Plan, Kerala, has been widely welcomed as an innovative participatory process capable of bringing the marginalized and weaker sections also in to the mainstream as empowered decision makers, rather than being outliers or passive beneficiaries. Its methodology is deeply embedded in empowerment of the people in the process of identification of development issues, formulation of projects, their monitoring and evaluation. The process also seeks the mobilization of voluntary labour, materials, expertise of serving and retired men in industry, agriculture, service sector, administration, teaching and research. Though the three-year period of People's Plan is too short for a meaningful evaluation of the process, our review of artisan-related projects has revealed significant gaps in conceptualization and planning. The scope for artisan project is brought out from our review of the development reports of many Grama Panchayats in Thiruvananthapuram district, where the number of rural artisans (17699) is the largest in the state. However, development projects for artisan beneficiaries have been formulated only in 23 Grama Panchayats; only half the projects (14 out of 28) have been successfully completed. All the projects address artisans as a whole and not traditional artisan as such, who are the worst victims of modernization, technology change, shortage of critical raw materials, instability in income and employment. Lack of entrepreneurial talent, leadership, capital, education etc. keep most of them as wage earners. Despite their skill, the traditional artisans are paid only little above the average wage of their unskilled counterparts in construction / agriculture. Non-traditional workers are sharing their traditional domains of work and there is very little incentive to keep them as skilled labour in the traditional sector or for the upgradation of skills through which they could be made fit for the emerging technology and environment.

In the light of the above facts, the tendency in the socio economic environment of traditional artisans and the gap in the artisan-related projects of Panchayats, the PRA workshop, which in fact turned out to be a 'grama sabha' of traditional artisans, proposed an 'artisan component' for the development plans of gramapanchayats with traditional artisan clusters. Participants of the workshop (mainly traditional carpenters) shared the view that a co-operative form of enterprise (wood/iron based building materials production co-operative) is necessary for a sustainable improvement in the situation facing traditional artisans in the local economy. Panchayats have been identified and acknowledged as vibrant agencies for the promotion and sustenance of such co-operatives of traditional artisans. Such participatory initiatives are envisaged as a beginning towards an 'Artisan Plan' on the model of Special Component and Tribal Sub-Plan.

REFERENCES

Development Report (1997-2001): Grama Panchayats in Thiruvanathapuram District.

Frances and Bee, Ronald (1998): *Facilitation Skills*, Institute of Personnel and Development, United Kingdom, Appendix-D: Force-Field Analysis.

Government of Kerala (1979): Committee for Artisan Development, Project Report.

Government of Kerala, Department of Economics and Statistics (1995): Report of Wage Structure-1982-92.

Government of Kerala, Department of Economics and Statistics (1996): Panchayath Level Statistics, Thiruvananthapuram District.

Government of Kerala, State Planning Board(1998 & 1999): Economic Review.

Issac, Thomas T.M and Harilal K.N (1997): "Planning for Empowerment: People's Campaign for Decentralized Planning in Kerala", *Economic and Political Weekly,* Volume XXXII, Nos. 1 and 2, January 4-11.

Kumar, Somesh (1999): Applications of Force Field Analysis in PRA, *Exchanges,* Issue Nos. 25 & 26, June and September.

Lewin, Kurt (1951): *Field Theory in Social Science*, Harper and Row, New York.

National Productivity Council (1979): *Rural Artisans in Payyannur: Report of a Study Sponsored by the Inter Agency Co-ordination Committee for Decentralized Sector*.

Reserve Bank of India (1974): All India Debt and Investment Survey 1971-72.

Sankaran, P.N (1996): "Institutional Reforms for Panchayati Raj: Towards Linking Cooperatives with Local Governments" in R. Rajagopalan [ed] *Rediscovering Cooperation*, Volume. I, Bases of Cooperation, Institute of Rural Management, Anand, Gujarat.

Sankaran, P.N (1997): "Panchayath Raj: Institutional Dimensions and Deficiencies", *Kurukshetra*, Volume XLV, No. 9, June.

Sankaran, P.N (1998): "Decentralized Planning: Opportunities for NGOs", *Employment News*, Volume XXIII, No. 28 10-16 October.

Sankaran, P.N (1999): "Empowerment and Welfare of Artisans: The Kerala Experience", *Kurukshetra*, Volume 48, No. 2, November.

6

PARTICIPATORY JUSTICE AND HUMAN RIGHTS PROTECTION IN LITTLE DEMOCRACIES

INTRODUCTION

Human rights is a twentieth century name for what has been traditionally known as natural rights or, in a more exhilarating phrase, the rights of man [Cranston, Maurice 1973]. The notion of natural rights has continued to attract man's minds; and the constitution or the legal codes of practically every state in the world today give at least formal recognition to the rights of man and the citizen. Thus one of the greatest accomplishments of civilization in the twentieth century is the acceptance of human rights as an integral part of governance for which the state, whatever its form, has accepted obligations and accountability [Madhava Menon, N.R 2000]. There is a strong NGO movement supported by an increasingly activist judiciary and an independent media to monitor human rights violations and to seek accountability of the powers-that-be at different levels of government. Admittedly, there is a human rights culture emerging everywhere informing and illuminating public policy and governance with varying degrees of success against strong currents of authoritarianism and abuse of power [Madhava Menon, N.R 2000 a].

Some studies have shown that poverty worldwide including India, has increased as a result of globalisation. According to the results of the 53rd round of the National Sample Survey released recently, the percentage of rural poor in India increased from 35% in 1991 to 38.5% in 1997. A survey carried out by the National Council of Applied Economic Research along with the International Development Research Centre of Canada drew similar conclusions. Yet another study by the International Food Policy Research Institute notes that while overall economic growth has been impressive since the start of reforms in the early 1990s, this has not made a positive impact on rural poverty. If after ten years of experience in economic liberalisation, there is increase in poverty and decrease in government's intervention in programmes directed towards alleviating poverty, human rights activists will have to ask questions of state accountability according to constitutional prescriptions. Human rights being part of the basic structure, no market-friendly government can hope to abridge it through constitutional review or similar exercise .... Can the state disown or disregard its obligations under the cover of economic liberalisation? [Madhava Menon, N.R 2000]. Similarly, shifts in economic policies under globalisation and the growth of Transnational Corporations pose new threats to human rights worldwide. When the state loses its capacity to decide its own economic policy, the first casualty is social welfare and thereby socio-economic rights..................The increasing power of Transnational Corporations and the inability of governments to control their activities lead to human rights violations [Koshy, Ninan 2002]. It is true that after the adoption of the Universal Declaration of Human Rights in 1948 (see Rights listed in Annex-I), an elaborate system of mechanism has been created to ensure the advancement of universal human rights, deal with violations of human rights, and to promote education/training in human rights. The emerging environment of globalisation and shrinking state, however, demand new strategies for human rights protection and promotion.

THE CONTEXT

In order to facilitate the promotion of human rights effectively, education, training and awareness programmes have been assigned a crucial role. The People's Decade for Human Rights Education, 1991-2001 [PDHRE] was therefore conceived in the late 1980's and a number of activities were carried out for the promotion of human rights education in many parts of the world. An International Congress on the Teaching of Human Rights was held in 1988 in Vienna. The UN World Conference on Human Rights [Vienna, 1993] Declaration inspired the UN Decade for Human Rights Education 1995-2004. Out of the 100 clauses in the Programme of Action of Vienna Declaration, four [78-81] are related directly to education and training. The World Conference on Human Rights [Programme of Action 78] considers human rights education training and public information essential for the promotion and achievement of stable and harmonious relations among communities for mutual understanding, tolerance and peace. Paragraph 33 of the Vienna Declaration and Programme of Action "emphasises the importance of .... human rights education...................... and calls on all states to do so". The section on human rights education in the Programme of Action recommends that states develop specific programmes and strategies of ensuring the widest human rights education, establishing international public policy on human rights education and reaffirming its importance.

HUMAN RIGHTS EDUCATION

According to Koenig, "the pursuit of human rights education serves as a powerful incentive for people to organise and coordinate themselves for effective action. The relationship between an active awareness of human rights and the realisation of these rights has been acknowledged worldwide" [Koenig, Shulamith 1994]. The purpose of mass education in human rights, he says, is to cultivate learning about justice, develop a universal human rights culture and create a global community in which abuse and violations are prevented and human dignity is protected by the powers of knowledge. In order to cultivate a human rights worldview, educators and grassroots activists understand that public education in human rights must demystify the laws that govern people's immediate lives, relate international standards-and more fundamentally the spirit and content of the Universal Declaration of Human Rights-to people's daily concerns. This learning must draw from spiritual and cultural experiences to be acceptable and meaningful. Human rights will thus be seen as relevant to people's relationship with each other as well as with the state and other entities (Koenig,Shulamith 1994). The fiftieth anniversary of the Universal Declaration of of Human Rights (1948-1998) with the theme "All Human Rights for All" -which fell during the Decade for Human Rights Education 1995-2004, provided an occasion for review of strategies for education, training and public awareness. In the scheme of human rights education, increasing involvement of international agencies, national/state level Commissions, NGOs, Voluntary organisations etc. is envisaged/welcomed.

THE INDIAN SCENARIO

In India the National Human Rights Commission (NHRC) set up in 1993 is expected to play a leading role in the task of human rights protection, education and training. Steps have been initiated to introduce human rights as a separate subject in schools and universities with the active involvement of the National Council of Education Research and Training (NCERT) and the University Grants Commission (UGC) in the process. But there is a view that the NHRC has not been truly involving human rights activists, NGOs etc. in the development of its approach to human rights education and training (Pal, R.M 1999). (See Annex-2 for a list of NGOs in the rolls of the Kerala State Human Rights Commission). Given the scope of education and training in the protection and promotion of human rights, and the threats posed to them by liberalisation/globalisation, there is need for developing micro-level strategies to supplement the efforts of the NHRC and State level Commissions at the macro level. As already stated, the ongoing marketisation and globalisation involve a gradual withdrawal of the state from welfare functions, posing grave threats to their obligations for the protection of human rights in general and socio-economic rights in particular. The emerging environment, therefore demands more emphasis on prevention rather than solutions in respect of human rights violations. Participatory approaches to prevention-focused/oriented education, training and awareness, through Panchayati Raj Institutions (PRIs)- the little democracies at the grassroots-therefore, need a fair trial. People's institutions could be strengthened to serve as a watchdog against micro level violations of human rights.

Though Panchayati Raj and Nagarapalika institutions have come to stay, following the 73rd and 74th Amendments to the Constitution of India, the question whether these participatory institutions could serve as a vibrant mechanism for the promotion/protection of human rights, especially the socio-economic rights, has not been seriously addressed so far. It may be remembered that the PRIs of India have a tradition of serving as local courts known as 'Nyaya Panchayats'(NPs).

In view of the present status of PRIs, with timely elections, functional and financial autonomy/stability, it is hypothesised that these institutions would be able to serve as an effective forum for human rights protection education and training in the new millennium. The scope and challenges of the proposed initiative is worth examining.

PARTICIPATORY JUSTICE: THE INDIAN TRADITION

Difficulty in access to justice lies in cost/delay in the administration of the judicial system so far as masses of Indian people are concerned and even classes of India are concerned. Therefore the point in relation to NPs is that the only way we can provide access to justice to rural India is by restoring the process known to India for thousands of years: that is the process of participatory justice where self determination exist in doing of justice by the people concerned (Upendra, Baxi 1986). In many parts of India, village panchayats constituted a vibrant forum of decentralised rural administration. Despite their limited geographical jurisdiction, the village panchayats were able to widen the range of their operational jurisdiction over the years, covering in a small measure judicial activities also. The institution of village panchayats continued to serve as the people's forum in diverse backgrounds and operated with informal mechanisms for improving social, economic, cultural, judicial and educational conditions of the village folk. For models of participatory justice we have the cases of Nagas, Bhils, Raj Gonds or the entire tribal groups in India. They have very sophisticated methods of determining the most complex disputes, as complex as those which come to the Supreme Court. This tribal traditional panchayats have very complex issues coming up for deliberation and decision. They are thoroughly de-institutionalized. They do not have the power of the kind that the state law has. Nevertheless they have a whole variety of sanctions and sanctioning mechanisms that have worked for centuries (Upendra, Baxi 1986). Baxi is not in favour of reviving NPs as such for the provision of participatory justice, because the NPs have functioned essentially as the assaulting arm of state power on local institutions. They displace traditional forms of dispute settlement like the Bardari Panchayat, the caste panchayat or panchayat dispute-settlement procedures developed by social agencies, voluntary organisations and social activists in certain areas. Hence he has suggested that the entire notion of NPs has to cease to be a mechanism for decentralisation and delegation of judicial power and has to become an instrument for organising the tradition and talent of Indian people for dispensation of justice.

In South India, the erstwhile Travancore and Cochin states and Malabar provinces of the former Madras State also had instituted Village Panchayats and Courts with limited pecuniary and other jurisdiction. Despite the useful functions performed by the village court, they were driven out under the influx of British common law and the system of judicial administration that followed. Panchayat courts as a chief, people-friendly and speedy forum for dispute redressal gradually disappeared; the fate of the village panchayat was also not much different. Through a series of state initiatives (frequently marked by setbacks), we have been able to rehabilitate the PRI system over the years especially in the 1990s, thanks to the 73rd and 74th Amendments to the Constitution of India. Though the developmental issues and socio-economic environment have changed in independent India, particularly with the advent of economic reforms, the village communities are not completely free from human rights violations and in most cases the affected parties are completely ignorant about their rights too. The need of the hour is to create awareness about various human rights and devise cost-effective methods to bring justice without delay.

In view of the fresh lease of life of the PRIs, following the passage of the 73rd and 74th Amendments to the Constitution, it is hypothesised that the village courts would be able to serve as an effective forum for judicial intervention. Closely connected with the revival of the traditional functions of village courts, the scope and challenges of human rights protection, promotion, training and education through the PRIs can also be looked into. Human rights intervention works towards the realisation of PRI objective of social justice. It may be remembered that a number of NGOs, churches, *Jagratha Samithies* (JSs) etc operate in the process of decentralised conflict resolution and protection of human rights. Hence the empowerment of PRIs through village courts or as designated agencies for human rights protection, promotion, education and training would sharpen their teeth and help realisation of the objectives of social justice. As stated above, the scope and challenges of the proposed initiative is worth examining.

OBJECTIVES OF THE STUDY

The objectives of the study are the following:

(a) To trace the evolution of Village Courts in Kerala and examine their linkages, if any, with human rights issues

(b) To identify formal and informal non-governmental mechanisms/organisations, if any, in operation for decentralised system of dispute redressal/conflict resolution/human rights protection at Panchayat level

(c) To find out cases of human rights violations in rural settings, and

(d) To explore the scope and challenges of Nyaya Panchayats and intervention of PRIs as agencies for the advancement of human rights at the grassroots level.

METHODOLOGY

The present study makes use of both secondary and primary data/information. The former has been collected from the Administration Reports and related publications of Travancore, Cochin and Malabar. To identify non-governmental agencies operating in the field of decentralised conflict resolution, human rights protection, dispute settlement etc we have undertaken a study of Pothencode Grama Panchayat in Thiruvananthapuram district, with special reference to Vengode, Ward 2 of the above panchayat. For the selection of the panchayat (and for intensive study, the Vengode Ward), the coverage of human rights organisations, non-governmental organisations, heavy concentration of backward classes particularly

SC/STs etc were taken into account. A questionnaire was developed making use of the insights from preliminary discussions/interaction with a human rights organisation in the study area. Opinion, suggestions and views of elected and non-elected persons in the panchayat, members of SHGs, representatives of communal organisations, groups of men and women were canvassed with the help of the questionnaire. Simple statistical tools like percentage and average have been used in the processing and interpretation of data.

NYAYA PANCHAYATS IN RETROSPECT AND PROSPECT

It is common knowledge that a litigation, civil or criminal, whichever party wins or loses, leaves a trail of bitterness and bad blood and tremendous financial loss for both, apart from protracted trials and resulting frustration. In Indian conditions, law and justice at the Panchayat level with a conciliatory methodology is an imperative necessity. This can be achieved only by having a suitable forum for conciliation and adjudication involving little cost and no delay with an informal procedure conforming only to the requirements of natural justice where the key word would be justice rather than law (Mathur, S.N 1997). Decentralised justice in the above background found expression in the concept of Nyaya Panchayats incorporated in the panchayat legislation of some states in the past. The Constitutional Amendments that gave birth to the new panchayats were, however, completely silent on the subject. Hence the question of decentralised justice now needs to be vigorously reviewed.

NYAYA PANCHAYATS, PEOPLES' JUSTICE AND PROFESSIONAL JUSTICE: A DISCUSSION

The institutional concept of Panchayats and 'Nyaya Panchayats' (peoples' justice) goes back to the hoary past not only in our country but in other ancient civilisations also. The judicial function of Panchayats is not a newfanged populist or proletarian innovation; it was perhaps the primary functions of the 'Panchayats' of 'jati', 'gram' or 'janpada' from times immemorial. Reference to Nyaya Panchayats abounds in ancient literature and historical documents. They were creations of villages themselves and were composed of persons who were generally respected and to whose decisions the villagers were accustomed to unqualified obedience. "Sitting on the Panchayats, the elders of the village used to solve disputes arising between the members of the village community. .... They heard parties in the presence of the entire village and solved the disputes. Public opinion of the village used to act as a powerful influence on the parties to the disputes and because justice was meted out at the very place where the dispute took place, it used to be inexpensive and immediate" (Report of the Committee set up by the Congress Working Committee, 1954). NPs were village self-governments in action at the level of administration of justice; it helped in involvement of people in dealing with law and justice and institutionalization of expeditious local resolution of disputes (Mathur, 1997). Mahatma Gandhi was the pioneer to revive the system of village justice when he advocated boycott of professional justice. NPs would be able to get over many of the limitations of the administration of professional justice. They are manned by people fully aware of local customs and habits; NPs follow low cost informal procedure, free from technicalities, and are closer to the people geographically and psychologically, so that values and attitudes are shared and there exists greater scope for conciliation/compromise. The following specific advantages are claimed for 'Nyaya Panchayats' viz; [1] they promote accessibility for the poor people to the institution of justice and help those who are priced out of the judicial market, [2] there is informality of the procedure in the Nyaya Panchayats; most of the disputes are amicably settled, [3] Nyaya Panchayats provide a forum where cheap and speedy justice is made available, [4] they are the carriers of secular, equalitarian modernalistic legal ideology, and thus, assist the desired social transformation (Baxi, Upendra 1980), and [5] to some extent they help in reducing the arrears of the court litigation.

NYAYA PANCHAYATS: A HISTORICAL REVIEW

Philosopher-theologist, Maharishi Nárad and Yagnavalkya, makes reference to traditional courts of ancient India. The king seldom interfered in the affairs of the panchayats and used to act as their guardian. Members of the traditional courts were well versed in *Vedicdharma* and were men of high integrity. Kautilya in Arthashastra also mentions these panchayats.

Village panchayats in the country continued to function relatively undisturbed during the Muslim rule. However, they suffered decline in jurisdiction and authority with the downfall of the Mughal rule. The British system of justice almost stifled peoples' participation in the administration of justice except in caste and revenue matters. Even under the British rule, few provinces like Madras, tried to rehabilitate traditional courts. Nevertheless, village panchayats and NPs fell prey to official indifference and in most provinces became defunct, despite the suggestion of the Royal Commission on Decentralisation that 'Panchayats should have civil and criminal jurisdiction in petty cases arising within the village'. The Indian Police Commission (1902), Government of India Act (1915), Bombay Village Panchayat Act (1920), the Civil Justice Committee (1924-25) and the Bombay Village Panchayat Act (1937) were in favour of NPs at the local level.

After independence also several attempts were made to revive NPs as part of strengthening panchayats on the basis of the Directive Principles of State Policy of the Constitution. Balwantrai Mehta Committee (1957), Ashok Mehta Committee (1978), the 14th Law Commission and the Rajagopaul Study Team (1962), also made valuable recommendations for the rehabilitation of NPs in the country. Although NPs were constituted in 15 states by special legislation, only few states actively promoted them. Reports of various evaluation committees appointed by state governments presented a discouraging picture.

A number of experts have examined the functioning of NPs in the context of Uttarpradesh,(R. Kushawaha, 1977, Richard, n.d), Rajasthan (Mathur, Iqbal Narain and Sinha, N.D) and Karnataka (Upendra Baxi, 1986). The general conclusions of the studies are the following:

- NPs have done some good work, but their potential remain untapped
- Their constitution has been by nomination or a combination of election with nomination
- There was a preference for NPs presided over by retired judicial officers or by new recruits for the purpose
- Lay members were subject to local influence and pressures; they required training
- Limited jurisdiction reduced their effectiveness
- Many panchayats were not adequately educated to discharge their functions effectively
- Meetings of NPs were irregular and their finances were also poor, and
- Caste, factions, stratification and local politics created difficulties for NPs in impartial operations.

According to Mathur, the most important factor that emerges out of these studies is that the political will and political awareness of the people are crucial for making the process of democratic decentralisation including the institution of NPs successful. The present scenario, as Alice Jacob observes, seems to be that Nyaya Panchayats have not established themselves as effective instruments for settlement of disputes.

NYAYA PANCHAYATS IN KERALA: AN OVERVIEW

(a) Travancore

Up to the year 986 (1810) there were no established courts in the country (Travancore). All civil matters then were heard and decided by *Surwadhikariakars* or by panchayats according to the order they received from the Maharaja or his minister whether he will be known by the name of *Valiyasurwadhy, Dalavah* or *Dewan*. They need to report the decision to the minister and according to the sanction accorded, execute the decree. Fees were also levied from the parties. (A Handbook of Travancore 1881).

With a view to reduce the number of petty suits in the civil courts, 19 Village Panchayat Courts were established in 19 Municipal Towns including Alleppey and Kottayam and their jurisdiction extended to money suits upto the value of Rs. 50. Village courts consisted of not less than 5 Honorary Judges, one of whom was styled the President. The government once in two years appointed them. Every suit had to be heard and disposed of by a Bench of atleast three Judges. Parties aggrieved by the decisions of the Panchayat Courts were entitled to have such decisions revised by the District Judge within whose jurisdiction the court was situated. The powers of the Village Panchayat Courts in regard to execution of decrees were limited to attachment of movable properties and in cases where other modes of execution were sought for, such as arrest of judgement-debtor and attachment and sale of immovable property, resort had to be had to the Munisiff's Courts (A Judicial History of Travancore, 1932).

(b) Cochin

In Cochin, Village Panchayats were in existence for a considerably long time. The Village Courts Act passed in 1943 repealed the major part of the Village Panchayats Act. The main work that the Village Panchayats were expected to discharge was judicial and with the constitution of separate Village Courts, the Panchayats were relieved of a large part of their work. Their services were later utilised to further and improve the activities of government in rural areas (Cochin Information, 1944). It was for the speedy and efficient administration of justice in the rural parts that the Panchayat courts were abolished and reconstituted as Village courts. Thirty-five village courts were constituted in 1118 immediately after the passing of the Village Courts Act and 12 in 1119. All the villages in the state were brought under the jurisdiction of the village courts. There was also provision in the budget for conference of the judges of the village courts (Cochin Information,1945).

(c) Malabar: Village Courts

Village headman (*adhikaris*) were empowered under the Village Courts Act of 1889 to try petty suits where the value of the cause of action does not exceed Rs. 20; and the system of trial by Bench Courts under section 9 of the same Act was in force in certain areas. But the Village Courts were not popular and did not appreciably lighten the burden of the civil courts. More than a quarter of the total number of suits launched in the Presidency in 1903 was filed in the Village courts; but in Malabar the proportion was only one-eleventh (Malabar Gazetteer).

Nyaya Panchayats (NPs) were functioning fairly satisfactorily in villages during the pre-British days. They were in fact the base on which the entire superstructure of indigenous system of administration of justice was founded. But on the coming of the British rule, an alien system of administration of justice was implemented in our country and Nyaya Panchayats, which were till then administering justice in rural areas, gradually lost their authority and influence and fell into disuse (Krishna Iyer, V.R 1979). At the time of commencement of the Constitution of India, some states (Madras, Mysore and Kerala) had the system of village courts whereas a few other states adopted it in pursuance of the Directive Principles of State Policy enshrined in Article 50 of the Constitution. Article 40 of the Constitution of India enjoins the state to organise village panchayats. The justification for 'Nyaya Panchayat's has been that they facilitate decentralisation and people's participation in one great branch of the government, namely, dispensation of justice (Jaswal, P.S 1996).

In spite of these laudable objectives, till now the attempts of some states to establish the NPs did not prove successful and some states (Maharashtra, Rajasthan) have gone even to the extent of abolishing them. Their failure was due to many reasons; they had only limited civil and criminal jurisdiction confined to petty offences without much power to enforce compliance/awards, issue directions etc. In many cases powerful landlords dominated NPs depriving common people the benefit of impartial justice. The NPs have a relatively insecure and subordinate status both as judicial institutions and as institutions operating within the context of Panchayati Raj system, without being an integral part of it.

Given the long tradition of 'professional justice', people's justice is an area of considerable controversy on which historians and political scientists may not readily agree. But "what we need is robust common sense and equally robust faith in the common man; without that faith in the capacity of the common man to govern himself and to learn to be fair in his dealings and in dispute situations, the Panchayati Raj experiment is foredoomed to failure. On the other hand, it would be unrealistic to assume the common man or his chosen representative to be inherently infallible or invariably just and fair and not to provide for their training or for other safeguards. A pragmatic middle path between "people's justice" and "professional justice" and a reconciliation between "democracy" and "rule of law" as understood and interpreted by lawyers and judges is not only possible but eminently practicable and desirable. Nyaya Panchayats could in course of time come to represent habits of self-government and respect for legality and justice which would be an enduring gain not only for grassroots democracy but also for grassroots rule of law (Singvi, L.M 1977). The tradition/ experience of the institution of Nyaya Panchayats in the resolution of local revenue, civil and criminal disputes would serve as solid base for commencement of a programme for human rights protection, promotion, training and education. There is in fact much wider scope for rethinking on such an intervention, given the constitutionally granted mandate and status of the present PRIs.

PANCHAYATI RAJ INSTITUTIONS, HUMAN RIGHTS AND DECENTRALISED SYSTEMS OF DISPUTE SETTLEMENT

(a) Panchayati Raj Institutions and Human Rights

The ongoing marketisation and globalisation involve a gradual withdrawal of the state from welfare functions, posing grave threats to its obligations for the protection of human rights in general and socio-economic rights in particular. As observed in the introduction, the emerging environment therefore demands (1) more emphasis on prevention rather than solution to human rights violations and (2) development of micro level strategies at local government levels to supplement the efforts of national and state level human rights commissions at the macro level. Participatory approaches to prevention-focused/oriented education, training and awareness, through Panchayati Raj Institutions (PRIs)–the little democracies at the grassroots-therefore, need a fair trial. Peoples' institutions could be strengthened to serve as a watchdog against micro level violations of human rights. Though PRIs have come to stay, following the 73rd and 74th Amendments to the Constitution of India, the question whether these participatory institutions could serve as a vibrant mechanism for the promotion/protection of human rights, especially the socio-economic rights, has not been adequately addressed so far, particularly in states like Kerala, where decentralised governance has become a model for other states. It may be remembered that the PRIs of India have a tradition of serving as local courts known as "*Nyaya Panchayats*". In view of the present status of PRIs, with timely elections, functional and financial autonomy/stability etc. it is hypothesised that these institutions would be able to serve as an effective forum for human rights protection, promotion, education and training in the new millennium. The scope and challenges of the proposed initiative is worth examining.

As a prelude to this, a brief review of the studies on PRIs and human rights is attempted below. In recent years, there has been a suggestion to link human rights protection and enforcement with PRIs, the little democracies of India. According to Haragopal there are three arguments in favour of the position that the PRIs will be able to promote and guarantee human rights; [1] these institutions lead to decentralisation of power and take decision-making closer to the people [2] grassroots institutions buttress the democratic culture, and [3] these institutions in the ultimate, contribute to the redistribution of power (Haragopal, G 1997). The PRIs, unlike several other institutions, are closer to the life and living place of the people in terms of size and space. It is precisely the reason why these institutions deserve a serious trial. We notice all over the country and particularly at the grassroots level that the coercive apparatus is becoming more important than the development apparatus. This shift in focus necessitates the linking and strengthening of PRIs with human rights agenda. Subordination of the police apparatus to the PRIs would serve the cause of effective decentralisation and also human rights (Haragopal, G 1997). For instance a provision like informing the arrest of an individual to the local PRI can reduce the arbitrariness in the arrest and detention. The PRIs for that purpose should constitute subcommittees with powers to review arrests and act as counter power centers. It is also maintained that a vibrant 'Grama Sabha' can not only fortify democratic culture but promote human rights and provide a fuller content to their meaning and also practice.

Mahipal has put forward three arguments for PRI promotion of human rights viz; through decentralisation of power, inculcation of democratic culture at local level and changing the existing unequal social order into a more equitable one (Mahi Pal, 2000). Operational autonomy, participatory approaches, empowerment of weaker sections, bottom-up procedures, transparency in functions, accountability to the people etc characterising the PRIs facilitate the promotion of human rights by them in a mutually rewarding manner. Besides the formulation of plans for economic development, PRI interventions towards the promotion of 'social justice', empowerment of weaker sections like SC/STs, OBCs, women etc [involving issues of human rights, particularly socio-economic rights] are essential imperatives of decentralised governance. Recognition of the 'Right to Development' as a human right (1986) emphasising equal access to basic resources, gender equality, removal of social injustice and participatory governance provide greater scope for PRI participation in human rights promotion. Through the process of judicial activism, during the last two decades, socio-economic rights viz; right to education, health, work, clean environment and development, reflecting the spirit of the Directive Principles of State Policy, are judicially declared as enforceable fundamental rights.

As they are also the basic concerns of the local governments or subject to the influence of their operations, the PRIs would in fact be strengthened through promotional human rights intervention under their patronage. For the advancement of socio-economic rights [which are labeled as Second Generation Human Rights], the PRIs would be an appropriate forum.

Studies conducted in different countries show that there is a strong linkage between decentralisation and human rights. While decentralised governance can promote respect for human rights in certain conditions, in others it may diminish civil status of a person or a group. International Council of Human Rights Policy (Geneva) has addressed these issues in the study 'Local Rule: Centralisation and Human Rights' (Edited excerpts, ISS, 2002). According to the study decentralisation might improve protection of human rights:

- When it enhances political rights (the right to take part in the government of one's country)
- When it leads to more effective government (transparency and accountability)
- When it helps achieve economic and social rights (efficient delivery of services–housing, education, health care etc.)
- When it promotes accountability (through informed participation of people), and
- When it increases local autonomy and empowers disadvantaged groups (through reservation of seats/office in local governments as in India for SC/STs, women).

There are certain conditions that might help decentralization of power to undermine the protection of human rights:

- When local populations are disempowered
- When national institutions are disempowered (for eg. professional associations, trade unions etc get fragmented, lowered in standards as in training and more exposed to corruption)
- When decentralisation exacerbates social divisions (special advantages available to minorities promote further fragmentation/duplication, separatism, conflicts, pressure groups etc.)
- When it deepens inequality (between regions and communities)
- When economic and social rights regress (through falling standards in services, declining devolution of finances from the central government, loss of economies of scale and lack of professionalism)
- While decentralisation helps the people living at the bottom of the society to assert their democratic rights, local elites fight back to retain their privileges, giving rise to social tension and violence
- When decentralisation provokes violence (in elections as in India, when newly elected councilors are beaten up, and killed by upper castes)
- When it empowers local elites (leading to local authoritarianism).

According to the study, central government clearly has an essential responsibility in relation to abuses of local power. Some of the ways it can intervene without undermining the decentralisation process are regulation, audit, investigation of allegations of abuse (by officials, police etc.) and development of institutions that monitor human rights (Ombudsman, Human Rights Commissions etc).

In a recent paper, George Mathew has examined the question whether the decentralisation process and the decentralised institutions increase human rights violations or enhance the possibility for respecting and observing human rights. The paper investigates first the social factors as well as denial of the right to livelihood, then those who are at the receiving end and finally the response of the state. According to him, there has been a sharp increase in violent manifestations of casteism in local communities ever since the local government systems got strengthened through the constitutional amendments. When the PRIs have been seen by the upper castes as the tool for the lower castes, to assert their right as individuals living in a democratic polity, the latter have become targets of caste-based discrimination and violence (Mathew, George 2003). Atrocities and human rights violations against the lower castes can be controlled only through a culture of equality of human beings. This can be achieved by human rights education and eradication of illiteracy. The Panchayats have a crucial role in this. Secondly, there are few states like West Bengal, Karnataka and Kerala where the issue of poverty has increasingly become a human rights issue and decentralisation is seen as an effective tool to address this problem. However, in many States, the poor remains where they are and in some cases their condition deteriorates, becoming more vulnerable to human rights violations due to corruption at the local level. Similarly after the elections, reports from States showed that the human rights of the dalits were violated in more than one sense even after strengthening of the local bodies. Violations of human rights are everyday occurance in the tribal areas of India, inspite of powerful legislations for decentralised governance. Reports from the length and breadth of the country indicates that women members of the Panchayats (from reserved constituencies and otherwise) have enhanced their status and rights. The rights of the people to participate in the decision making process of their locality is legally mandated through the Grama Sabha. However, frequent postponement of elections has made the above position a mockery. The creation of parallel bodies usherping into the domain of local bodies is another instance of delimiting the powers of the latter.

Against the background of the emerging role of PRIs in the field of human rights, a number of initiatives have been set in motion in the above direction. Tamil Nadu government has decided to form watchdog committees at village level for elimination of trafficking in women and children. The action was taken on the basis of advice given by state level Advisory Committee on Eradication of Child Prostitution. Each watchdog committee will consist of village panchayat (President), village administrative secretary (Convener), headmaster of the high school, a representative of the local police station, a representative of NGOs working in the area and a member of SHG active in the area (ISS, Panchayati Raj Update 2002).

In a recent study (reported by Raghaviah), Mohana kumar and Binitha (2002) have explored the scope and working of *'Jagratha Samithies'* in Mulamkunnathukavu, Trissur district, Kerala. According to them, in Kerala, where complaints by women relating to sexual harassment, marital frauds and the like are on the rise, panchayat level *Jagratha Samithies (JSs)* could be the answer to simplify the process of investigation and provide quick legal redress to women. *Jagratha Samithies* were thought of at a juncture when the size of the State Women's Commission (and by implication the volume of cases it can manage) was reduced to that of a nominal agency. *Jagratha Samithies* (JSs) have assumed the role of an informal dispute-investigating forum and has been acknowledged as an ideal means for reaching quick legal aid to women. Interestingly in Grama Panchayats where the JSs function, they have proved to be highly effective in investigating the grievance of women at the grassroots. A JS consists of 15 members (of which 2/3 are women) and 3 non-official members-doctor,. police official or lawyer, preferably women again. The nine member district level committee has as members the District Collector (Chairman), District Panchayat President, Municipal Chairman, Superintendent of Police and womens organisation representatives. The Grama Panchayats popularise the concept of JSs and provide for its expenses under their women component plans. The samithies facilitate timely reporting of investigations and redressal of atrocities against women through agencies like the police, family courts, Women's Commission etc. Despite the claims of the Kerala model, status of women in the state leaves much to be desired; their work participation is very low (16% as against 23% for India), only 6 per cent of them are represented in the State Assembly. According to a survey of the State Planning Board, while the number of incidents of violence against women has been recorded as 36.90 for the country, the figures are 62.30 and 68.30 for Thiruvananthapuram city and rural respectively.

From the survey of studies on PRIs' role in human rights protection, we find that there is much scope for such interventions in future. However, the space for such intervention by the PRIs needs to be properly defined and effective linkages have to be established with NGOs and other agencies in the field. It is against the above setting that we have undertaken the present study.

(b) Decentralised Systems of Dispute Settlement

It is a fact that a large number of civil and criminal cases, as also others involving human rights violations, are not taken to the formal system (judiciary) of the justice process. Several cases remain unattended/unreported/unsettled while others are settled through the informal systems of dispute settlement/conflict resolution. With the decline of 'Nyaya Panchayats', informal systems of dispute settlement have emerged in several parts of the country. In Kerala, apart from voluntary agencies, a number of communal organisations have diversified their operations into dispute settlement and conflict resolution. Various denominations of the Christian Church,

Muslim Organisations, Hindu religious outfits, communal organisations of Ezhavas, Nairs, SC/STs, Other Backward Classes etc. are actively involved (suo moto or upon complaints) in the identification and settlement of inter community as well as intra-community disputes relating to marriage, property, place of worship, employment, education etc. In fact such systems of informal and decentralised dispute settlement relieve the burden of the formal system, which is saddled with mounting backlog of undisposed cases involving high social cost over and above the personal cost of the complainants/petitioners. In many cases, it is the high cost, poor access and delay associated with the formal system, that draws the poor and the ignorant closer to the informal system of dispute settlement/rights protection. Whereas the formal system, like many government departments, does not provide much extension/promotional service in the delivery of justice to the affected people, the basic mode of operation of the informal system is the delivery of services through suo moto identification, negotiation and settlements, many a time through the participatory process. Though the informal system lacks power to enforce attendance of parties to an issue, call for documents or search for material evidence, the natural tendency of people to remain within the community-fold renders social control over them by the communal organisations. The participatory process is thus people-friendly, speedy, cost-effective and negotiation-driven. The informal system renders sustainable services to individuals and the society at large, by settling marital issues, discrimination against the poor, children, women, illiterate people etc, property disputes, religious issues and so on.

In spite of the awareness of the people about the limitations of the formal justice process and the efficacy of the participatory justice process, the role of the latter has not been adequately recognised, documented and empirically examined. The issue has assumed a new dimension in the context of the widespread acceptance of participatory interventions in socio-economic issues under the Panchayati Raj system, in which Kerala has been a pioneer in many respects. It is therefore presumed that participatory interventions in the justice process need close examination and proper linkage with the general framework of participatory socio-economic development, in the special context of Kerala. The conceptual framework of the present study thus owes very much to the ongoing attempts towards enlarging the participatory process, both in theory and practice, so that development is made sustainable and truly human.

MODELS OF DECENTRALISED/INFORMAL DISPUTE SETTLEMENT

(a) Oorukkoottom

Oorukkoottom generally refers to the village assembly or grama sabha. They were councils of village government in traditional Kerala which decided public affairs of the village including temple issues. They existed until the first quarter of the 19th century. A widely known decentralised system of (informal) dispute settlement, popular among the scheduled tribes of Kerala has been *'Oorukkoottom'*. It serves as a forum headed by elder members of scheduled tribes of Kerala inhabiting hilly areas of Trivandrum, Idduki, Wyanad and Kozhikode districts. *'Oorukkoottom'*, considered as a tribal assembly, settles disputes of all kinds involving the tribals inter se, and the decisions of the tribal leaders are accepted by all without going into the merits of such decisions. In addition to imposition of penalty, appropriate strictures are also passed in the form of warning and in certain cases, ostracisation of the disobedient/rebels from the tribal mode of living. Persons found to be guilty of grave offences are forbidden from participation in functions/ ceremonies including funeral, marriage, cohabitation, festivals, rituals, sharing food and related cultural practices. It may however be noted that, the mode of operation, procedures and the principles that are followed in such assemblies are purely caste-specific, deeply rooted in their own cultural milieu and are rarely in tune with or justified/tempered by the formal laws of the land. Despite this drawback, members of the tribes prefer to be governed by the law of the tribe rather than the common law of the land. In the above context, policy decisions on promoting the informal dispute settlement mechanisms have to pay serious attention to existing traditional and caste-specific mechanisms for dispute settlement before attempting structural changes in the system.

(b) Kadakkodi (Sea-Court)

'Kadakkodi' is a colloquial term coined from two Malayalam words *'Kadal'* (stands for sea) and *'Kodathy'* (stands for court). The *kadakkodi*, indigenous judiciary at the village level, was formed according to religious beliefs and systems. This is a forum to settle all disputes in the village, including fishery regulations. Interestingly, *Kadakkodi* is more than a judiciary as it performs legislative and executive functions too. It enacts rules and regulations, enforces and executes them (Rajan, 2002). The *Kadakkodi* in central and north Kerala, South India, had a stronghold in the fishery management regime. It consists of a committee unanimously elected by the village community. *Kadakkodi* is comprised of two forms of committees- Committee of Hindu Priests (from fishing community) and Committee of Village People. The former known as *Achanmaar*, is elected for lifetime according to hereditary Hindu rituals. There are 12 priests of different grades with different functions. The grade depends on the *Tharavad* or family status. The structure of the committee varies between villages. The community elects the Committee of Village People annually during the time of temple festival. This indigenous court exercised social control through enactment and enforcement of regulations on fishing technologies, seasons, timing, fishing rights etc, which are binding (socially, not legally) for the community. The Court had been vested with powers to settle social and family issues too. But this Court in the fisheries sector, which once had a fisheries management regime, has almost declined now.

The acceptance of the community, the consensus in settling disputes, the participatory and decentralised processes are the merits of the Court. The Court acts upon petitions, suo moto,or based on public interest litigation. The *Achanmaar* (Judges of the Court) usually gather at the temple premises. If they receive petitions, a flag will be hoisted on the seashore where the general body is usually convened in the evenings. Both the parties are summoned and examined. Sometimes, the Court will appoint a Commission comprising of independent persons from the village to enquire into the matter in detail. On the report of the commission, the Judges pronounce the judgment and it is announced in the general body. The aggrieved parties will be given the chance of *Punar* (appeal). Those who violate the rules relating to fisheries will be awarded punishment (in the form of excommunication) or will be denied the rituals on the occasions of birth, marriage, death etc.

From the above account of the informal, decentralised and participatory systems of dispute settlement, it may be concluded that they are caste, community and location-specific and are therefore unsuitable for state-wide application. In a society moving towards secularism, accompanied by high rate of horizontal/ vertical, social and occupational mobility, the traditional caste–linked institutions of informal dispute settlement lacks popular replicability. The most suitable institutional alternative, therefore, lies in the promotion of human rights organisations which are well entrenched in the Universal Declaration of Human Rights (1948), the embodiment of the natural rights of people world-wide. In such programmes of human rights protection, promotion and education, there exists a meaningful space for the institutions closer to the people, like the Panchayats, for mutually beneficial and participatory operations.

HUMAN RIGHTS PROTECTION, PROMOTION AND EDUCATION BY PANCHAYATS: SCOPE AND CHALLENGES

In order to examine the awareness of people about human rights issues in general and violations of rights involving death due to poverty, dowry-related atrocities/ death, custodial torture/death, and others in police stations, a sample survey was carried out in Pothencode Grama Panchayat, Trivandrum district. For the purpose of the enquiry, a draft questionnaire was developed which was discussed with selected members of a human rights organisation having one year of presence in ward II Vengode of Pothencode Panchayat. The survey was carried out during the third week of April 2003. Apart from the sample of permanent residents of Vengode, the response of officials and elected members of the panchayat was also canvassed. The sample of permanent residents was selected from a small gathering facilitated by the HRO and activists associated with the SC/ST colony in Vengode. Those who volunteered were asked to respond to the questionnaire. Members of SHGs were contacted at their place of residence. Ward II of Vengode panchayat was purposively selected because of the high concentration of most backward and marginalised classes of SC/ST population in the area. The presence of women and child welfare activities including women empowerment programmes like Kudambasree, Neighbourhood Groups (NHGs), Self Help Groups (SHGs), human rights organisations etc were also taken into account in the choice of Vengode as the study area.

As a prelude to the recording of responses from the sample (100), a brief account of the nature of human rights, history of Nyaya Panchayats and the scope of decentralized dispute settlement/conflict resolution at panchayat level was given to them. With a view to obtain the opinion of different categories of people dealing with/subject to human rights issues, local governance, dispute settlement etc., elected and non-elected members of panchayat, poor women following group activity (members of self help groups), and members of communal organisations of the locality were covered in the study, in addition to a separate category of unorganised men and women. A profile of the study area is given in Table 1. It is observed that the female population is 51 per cent and the scheduled caste population is 10 per cent of the total population.

Table 1
Profile of the Study Area : Pothencode Grama Panchayat, Thiruvananthapuram District

District	:	Thiruvananthapuram
Block	:	Kazhakoottom
Location	:	North of Thiruvananthapuram Taluk
Classfication	:	Special Grade
Village	:	• Upper Thonnackal • Lower Thonnackal • Ayirooppara • Aandoorkonam
Area	:	20.85 sq. km.
No. of wards	:	13
Population	:	24,995 12,729 Female 12,266 Male 2,600 scheduled castes 1,305 Female 1,295 Male
Economic Activity	:	Agriculture - 60% of population Own cultivation 15% Trade and petty jobs 8% Govt. employees 5% Outside employment 5% Service sector 10%
Employment Status	:	Educated unemployment
Own funds	:	Rs. 25 lakhs P.A.
Educational Institutions	:	Govt. LPS (2) Govt UPS(2) Govt. HS (1)
Health care	:	PHC Ayurveda Dispensary Family Welfare Centre Mannara
Women and Child Welfare	:	25 Ankanavadies Karunya Special School
Women Empowerment	:	Kudumbasree, Neighbourhood Groups (NHGs) Problems of Marketing products of women units.

Source: Pothencode Grama Panchayat, Decentralised Planning, Xth Plan (2002-07).

SOCIO-ECONOMIC STATUS OF RESPONDENTS

The general socio-economic background of the 100 respondents (Panchayat officials 8, Panchayat members-8, members of Self Help Groups (women)-26, members of communal organisations-11, Men-34 and women-13) is summarized in Table 2.

There are 47 women and 53 men in the sample of 100 respondents. The community-wise distribution indicates 49 scheduled caste members, 5 scheduled tribe members, 28 other backward class members and 18 forward class members covered by the present study. The analysis of educational status of the respondents shows that the average educational qualification is tenth standard. The average age group of respondents is 31-50 years; most of them (54%) face unemployment and 23 per cent follow other small activities. Given the above composition of the respondents, it may be noted that the study gives enough weight for the views of women and backward communities (usually considered as weaker sections of the rural society) on issues affecting their day to day life involving human rights violation, protection and the role of Panchayats in dispute settlement/resolution.

AWARENESS ABOUT RIGHTS AND VIOLATIONS

The response of the sample groups on general awareness about human rights, death due to poverty, dowry-related atrocities, custodial torture, police atrocities etc. is analysed with the help of Table 3.

Because of the operations of human rights and communal organisations in the study area, a higher percentage of respondents (83) irrespective of categories, have reported general human rights awareness. However, their awareness is not that much high about HR protection in the study area. The gap in awareness between rights and protection leaves scope for HR interventions in the study area. The lower number of beneficiaries reporting help rendered by HR organisations further amplifies this. The rights to employment and livelihood, freedom from illegal police torture, interference with the right to married life (without dowry related atrocities, torture, interference in privacy etc.) etc are important human rights. In the above backdrop, the opinion of the respondents was invited on the above issues. Reports on custodial atrocities and death due to poverty are few; however 50 per cent of the respondents have stated their awareness about dowry-related atrocities and related HR violations. About 50 per cent of the respondents, especially men and women groups in the sample, are aware of HR violations by the police. Because of this, only few respondents (35%) are satisfied with police intervention in HR/dispute settlement. The above result points to the need for reforming the dealings of the police towards the public and human rights education among policemen.

NATURE OF HUMAN RIGHTS VIOLATION

Awareness of the respondents about human rights violations and the nature of violations reported by them are analysed with reference to Table 4.

Table 2
Distribution of Respondents according to Sex, Age, Community, Educational Status and Employment

Respondents	No.	Sex		Community				Educational Status							Employment						Age		
		M	F	Forward	OBC	SC	ST	Illiterate	<5 Std.	10th Std	PDC	Degree	P.G	Professional	Govt.	Agri.	Industry	Business	Others	Unemployment	18-30	31-50	>50
Panchayat Officials	8	5	3	3	5	0	0	0	0	1	2	1	3	1	8	0	0	0	0	0	3	4	1
Panchayat Members	8	4	4	5	2	1	0	0	0	5	1	2	0	0	0	1	0	0	4	3	1	5	2
Members of SGHs Women	26	0	26	5	5	15	1	1	8	10	4	2	0	1	0	2	1	0	2	21	10	12	4
Members of Communal Organisations	11	10	1	0	9	2	0	0	2	4	3	2	0	0	0	2	1	1	1	6	5	4	2
Men	34	34	0	1	5	24	4	0	12	17	5	0	0	0	1	3	0	2	12	16	9	18	7
Women	13	0	13	4	2	7	0	1	9	1	1	1	0	0	0	1	0	0	4	8	7	3	3
Total	100	53	47	18	28	49	5	2	31	38	16	8	3	2	9	9	2	3	23	54	35	46	19

Source: Field Survey.

Table 3

Distribution of Respondents According to Awareness: Human Rights, Death due to Poverty, Dowry Related Atrocities/death, Custodial Torture/death and HR Violation in Police Station

Respondent	*No. having general HR awareness*	*No. aware of HR protection in Panchayat*	*No. of beneficiaries of HR organizations*	*No. Aware of Death due to*			*No. aware of HR violations by the police*	*No. satisfied with police intervention HR/dispute settlement*
				Poverty	*Dowry related atrocities*	*Custodial atrocities*		
Panchayat Officials	8	7	1	0	5	0	6	0
Panchayat Members	8	7	0	0	2	0	2	5
Members of SHGs (women)	20	19	12	8	12	14	10	16
Members of Communal Organisations	11	2	2	0	4	0	3	2
Men	25	13	18	7	18	14	21	4
Women	11	9	6	4	9	5	8	8
Total	83	57	39	19	50	33	50	35

Source: Field Survey.

Table 4
Awareness of Human Rights in the Panchayat & Nature of Violation

Respondents	*No. Aware of*	*Nature of Violation**																						
		1	*2*	*3*	*4*	*5*	*6*	*7*	*8*	*9*	*10*	*11*	*12*	*13*	*14*	*15*	*16*	*17*	*18*	*19*	*20*	*21*	*22*	*23*	*Total*
Panchayat Officials	2	1	1	1	0	0	0	1	0	1	0	0	1	0	1	0	1	0	0	0	0	0	0	1	9
Panchayat Members	0	0	0	0	0	0	0	0	0	0	0	0	0	0	0	0	0	0	0	0	0	0	0	0	0
Member of SHGs (women)	18	9	5	1	1	0	3	6	0	4	2	0	3	1	0	0	4	6	5	0	2	0	0	0	52
Members of Communal Organisations	8	5	0	2	0	0	0	4	0	1	0	0	0	1	3	0	2	4	0	0	1	0	1	1	25
Men	29	4	8	5	3	2	3	15	2	12	2	2	3	0	3	2	7	18	10	4	0	1	1	1	108
Women	12	6	4	0	0	0	0	6	0	3	0	0	1	0	0	0	1	3	1	1	0	0	0	0	26
Total	69	25	18	9	4	2	6	32	2	21	4	2	8	2	7	2	15	31	16	5	3	1	2	3	220

Source: Field Survey.

* Human Rights Violation involving: 1. Women 2. Children 3. Poor 4. Illiterate 5. Political Parties 6. Religion 7. SC/STs 8. Atheists 9. Unemployed 10. Criminals 11. Ex-prisoners 12. Domestic Servants 13. Mixed Marriage 14. Widows 15. Separated 16. Dowry atrocities 17. Police atrocity 18. Custodial death 19. Election Candidates 20. Educational institution 21. Cultural institution 22. Place of work 23. Public property

Majority of the respondents (69%), except panchayat members, has expressed their awareness about human rights violations in the panchayat. The exclusive category of women and women operating under SHGs together (39) have expressed high level of awareness; it is observed that 77 per cent of them are aware of different types of HR violations in the study area. A high percentage of men (85%) have also reported their awareness. It may however be noted that the political leadership (elected members of the panchayat) is reluctant to accept the existence of HR violations in the panchayat.

The nature of HR violations and the category of respondents reporting awareness about different types of violations are also examined in the study. We have identified and examined 23 situations of HR violations involving men, women, children, poor, backward categories, criminals, election, educational/cultural institutions, place of work, public property, environment etc. An analysis of the violations reported (220) brings out highly informative results with implications for better human rights awareness programmes and protective measures. Some of the respondents have reported awareness about multiple types of HR violations in the locality. Men have stated 108 cases of HR violations, forming 50 per cent of the responses obtained in the survey. A careful scanning of the distribution of violations involving different types of people and situations provides interesting observations indicating the gravity of violations and the socio economic background of the victims. It is widely held that weaker sections especially SC/STs, women and those who happen to deal with the police machinery are the usual victims of HR violations in a developing society. Our study also endorses the prevailing viewpoint on HR violations against weaker sections and women in workplaces, households, police station, public places etc. It is noted that 15 per cent of the reported violations involve SCs and STs and 14 per cent of the cases are connected with police atrocities. Cases of custodial death (7%) reinforce the general findings stated above. Dowry-related atrocities (7%), reported in the study also need to be viewed seriously, as they also involve cases of exploitation of women, posing a grave threat to a very strong unit of the society-family.

ROLE OF ALTERNATIVE AGENCIES IN HR PROMOTION AND EDUCATION

At the global, national, regional and local levels the role of HR awareness programmes in HR promotion/education has been increasingly emphasised. Prevention is better than cure: the dictum holds true in the case of HR violations also. Human Rights Commissions, Educational institutions, NGOs and Civil society organsiations undertake HR promotion and education programmes. Against the above setting, the study collected information about HR awareness programmes, the agencies involved, the role of Grama Sabha if any, nature of dispute settlement, effectiveness of settlement, scope of HR education in panchayats etc. Majority (62%) of the respondents are aware of HR programmes in and around the study area. The study area, having the presence of an HR NGO, provides much scope for NGO-led intervention in HR awareness, promotion, education etc. Out of the reporting respondents (62%), about 79 per cent hold the view that HR organisations are instrumental in awareness, prevention and education. The panchayat is found to be contributing the least in the above matter; infact its role is observed to be the least in our survey covering four agencies, namely panchayat, HROs, COs and NGOs. Interestingly NGOs and COs involve in such programmes better than the panchayat.

The Grama Sabha is considered a potential (people's) forum for raising and settling local issues, disputes etc. Hence we have addressed the issue whether the Grama Sahba is utilized in the above perspective. Though 47 per cent of the respondents have reported cases of HR issues raised in Grama sabha, the forum is generally considered very weak for settlement of HR issues. However, the scope of an HR working group in Grama sabha is highlighted by 90 per cent of the respondents; members of communal organisations and SHGs are found to be the most powerful votaries of the suggested HR intervention through the Grama sabha.

In many parts of the country for sometime panchayats were involved in dispute settlement through the machinery of Nyaya Panchayats. As the panchayats of today have a solid foundation of timely elections, adequate funds, personnel and local participation of people in their programmes, they can think of reviving their old form of intervention in local dispute settlement. In the above setting, we addressed the question on the effectiveness of the panchayat in dispute settlement. Though the present level of intervention by the panchayat in dispute settlement is very low, majority (66%) of the respondents hold the view that panchayats can have a role in the matter (including HR issues). The potential role of the panchayat is supported fully by panchayat members followed by women SHG members and representatives of communal organisations (Table 5).

Our study has brought out a highly supportive role for the panchayat in HR education; 89 per cent of the respondents hold the above view. Elected members of the panchayat, women SHGs and members of communal organisations fully endorse the possible role of panchayat as a forum for HR education. Other categories of respondents like employees of panchayat, women groups are also supportive of the HR educational programme envisaged in the panchayat.

SCOPE AND CHALLENGES OF NYAYA PANCHAYATS

Along with HR issues in the panchayat, the scope of HR promotion and education through the panchayats and reviving another institutional intervention (that existed decades ago), viz Nyaya panchayat, is also addressed in our study. It is observed from Table 6 that 54 per cent of the respondents have heard of NPs. Among the different categories of respondents, all the members of the panchayat, followed by employees of panchayat and members of communal organisation have reported their knowledge about the old NP system.

Nearly 79 per cent has expressed awareness about 'Oorukuttom' (a traditional informal institutional setup for the settlement of various issues of SC/STs in their clusters). However, its effectiveness is not shared by the majority. It is found that 65 per cent of our sample welcomes NPs in the panchayat systems. Respondents

Table 5

Human Rights Awareness and Role of Different Agencies in its Promotion/Education

Respondents	*No Aware of programmes*	*Agencies involved**				*HR issues raised in GS*	*Whether Settled*	*Scope of HR working group in Grama Sabha*	*Effectiveness of Panchayat in settlement of disputes HR issues*	*Scope of HR education in panchayat*
		P	*HROs*	*Cos*	*NGOs*					
Panchayat Officials	2	1	0	0	1	3	1	7	5	7
Panchayat Members	6	0	5	0	1	2	2	7	8	8
Members of SHGs (Women)	17	1	16	0	0	14	6	25	19	26
Members of Communal Organisations	2	1	1	0	0	2	0	11	8	11
Men	23	0	16	3	4	16	10	28	18	26
Women	12	0	11	1	0	10	6	12	8	11
Total	62	3	49	4	6	47	25	90	66	89

Source: Field Survey

* P: Panchayat *HROs: Human Rights Organisations *COs: Communal Organisations. *NGOs: Non-Governmental Organisations.

Table 6

Opinion on the Scope and Challenges of Nyaya Panchayats (NPs)

Respondents	*No. heard of NPs*	*No. aware of Ooruk-kuttom*	*No. stating effectiveness of Ooruk-kuttom*	*No. wel-coming NPs in PR system*	*No. favouring NP in*		*No. welcoming an impartial NP*	*No. fearing political interference in NP*	*No. reporting positive effect of NPs on BCs, widows, women etc.*	*No. welcoming petitions to Grama Sabha to entertain NP*
					Old form	*With Changes*				
Panchayat Officials	6	8	6	5	1	7	8	8	7	4
Panchayat Members	8	8	4	5	1	7	8	4	8	8
Members of SHGs (women)	9	15	14	13	4	22	6	18	13	21
Members of Communal Organisations	8	8	3	9	0	11	11	10	10	11
Men	18	31	15	23	1	33	33	23	24	32
Women	5	9	7	10	4	9	11	5	10	13
Total	54	79	49	65	11	89	77	68	72	89

Source: Field Survey.

associated with the panchayat (62.5%) and women as a separate category (77%) are the groups that strongly welcome it. Opinion is not in favour of reviving the old form of NPs, but with suitable changes (89%). All the respondents, especially those linked to panchayat, welcome the NPs provided they are impartial and effective. Similarly the effect of NPs on BCs, widows, women etc. is reported by 72 per cent of the respondents as being positive. The forceful supporters of the view are members of the panchayat (100%), employees (88%), representatives of COs (91%) and women (77%).

Elections in India, down to the local level are fought on political basis and there has been progressive politicisation of public issues, particularly in Kerala. Eventhough the virtues of participatory interventions in local governance are widely shared, we have observed that 68 per cent of the respondents anticipate a high degree of political interference in the proposed NPs (employees of panchayat (100%, COs 91% , SHGs 69%). However, only 50 per cent of the members of the panchayat subscribe to the dominant view of politicisation of NPs. Majority of the respondents (89%), welcome the grama sabha as a forum for screening petitions to be placed in the proposed NPs. Elected members of the panchayat, communal organisations and women groups fully support (100%) the proposed role of the grama sabha in between the petitioners and the NPs.

PREFERRED AGENCIES FOR DISPUTE SETTLEMENT/RIGHTS PROMOTION

At the local level of panchayat, a number of agencies exist for the settlement of disputes/resolution of conflict, involving the family, community, place of worship, marriage, property (private & public), environment etc. Institutional preferences of the respondents are addressed in Table 7.

Police station, COs, HROs, Judiciary, voluntary organisations, panchayat, political parties and free legal aid agencies are the mechanisms posed for the expression of respondent's preferences. Being a pilot study, we have not delved deep into the order of preference in cases of multiple preferences (147 responses were obtained from 100 respondents, indicating more than one preference stated by few members). The most preferred (44%) dispute settlement / rights protection agency is HROs and the least preferred one (2.72%) is political parties. Free legal aid agency (11.56%), judiciary (10.88%) and police machinery (10.20%) are the other mechanisms favored, as may be noted from the responses of the sample groups. The distribution of responses indicate a preference for the panchayat (7.48%) compared to COs and other voluntary agencies (6.80 each). It may be noted that, despite the critical attitude to the police apparatus, respondents have an express preference for it over political parties, voluntary organizations and the panchayat, in matters of dispute settlement and local conflict resolution.

The findings and observations of the Pilot study may be summarised as follows. Though a high percentage of the respondents (83) are aware about human rights in general, the level of awareness is low with regard to protection of human rights in the study area. Human rights violations involving dowry-related atrocities, and others linked to the police machinery have been reported by 50 per cent of the respondents. There is a case for human rights education among policemen, since only 35 per cent of the respondents are satisfied with police intervention in human rights/dispute settlement.

Table 7
Agencies Preferred for Settlement of Disputes in Family, Community, Place of Worship, Marriage, Property, Environment etc

Respondents	*Agencies Preferred*							
	Police Station	*Community Organisation*	*Human Rights Organisation*	*Judiciary*	*Voluntary Organisation*	*Panchayat*	*Political parties*	*Free Legal Aid Agency*
Panchayat Officials	1	0	1	1	0	1	3	1
Panchayat Members	2	0	0	0	0	7	0	1
Members of SHGs (women)	4	0	25	4	3	1	0	4
Members of Communal Organisations	0	7	4	0	0	0	0	2
Men	6	2	24	8	6	1	1	9
Women	2	1	10	3	1	1	0	0
Total	15	10	64	16	10	11	4	17

Source: Field Survey.

Large number (220) of human rights violations reported by the respondents indicate the gravity of the situation existing in rural Kerala and the socio-economic background of the victims (mostly SC/STs, women and others dealing with the police machinery). About 15 per cent of the reported violations involve SC/STs and 14 per cent of the cases are connected with police atrocities. Out of the reporting respondents (62%), about 79 per cent hold the view that human rights organisations are instrumental in human rights awareness, prevention and education. It is also observed that NGOs and COs involve in such programmes better than the panchayat. Though Grama Sabha is considered a potentially useful forum for the human rights agenda, its intervention is rather modest; there is scope for the working group arrangement in Grama Sabha for human rights issues.

All categories of the respondents support the proposal for human rights education through the panchayat. Similarly 65 per cent of the sample welcomes 'Nyaya Panchayats' in the panchayat system. However, political interference is viewed as a major threat to the impartial operation of Nyaya panchayat; hence suitable safeguards will have to be instituted to make the proposed Nyaya panchayats impartial, corruption-free and people-friendly. Analysis of people's preference for institutional alternatives in dispute settlement/human rights protection has brought out a strong case for human rights organisations in villages, especially in those inhabited by backward classes/weaker sections. Inspite of the highly politicised environment existing in Kerala, covering elections as well to local self-governments, political parties are the least preferred agency in dispute settlement. Despite the critical attitude to the police apparatus, respondents have an express preference for it over political parties and panchayat, in local conflict resolution and dispute settlement.

The respondents have also made independent suggestions for protection of human rights and effective working of Nyaya Panchayats. Ward/panchayat level awareness programmes by HROs, COs, Panchayats and NGOs have been suggested by all the respondent groups. Similarly the need for a more detailed macro level survey/study on the state of human rights and introduction of a model NP on an experimental basis have also been suggested.

CONCLUSIONS AND SUGGESTIONS

Human rights are universal legal guarantees protecting individuals and groups against actions, which interfere with fundamental freedoms and human dignity. Some of the most important characteristics of human rights are the following:

• They are internationally guaranteed • They are legally protected • They focus on the dignity of human beings • They protect individuals and groups • They obligate states and state actors • They cannot be waived/taken away • They are equal and independent • They are universal. They include a broad range of guarantees encompassing every aspect of human life and interaction. The popular rights are • The right to life • Freedom from torture and cruel, inhuman or degrading treatment or punishment • Freedom from arbitrary arrest or detention • The right to fair trial • Freedom from discrimination • The right to equal protection of the law • Freedom from arbitrary interference with privacy, family, home or correspondence • Freedom of association, expression, assembly and movement • The right to vote and take part in government • The right to just and favourable work conditions • The right to adequate food, shelter, clothing and social security • The right to health, education and property • The right to development. The implementation of human rights standard is closely watched at several levels like government agencies and services, human rights commissions, ombudsman, HROs,NGOs, community-based organisations, the courts, parliament, the media, professional associations, trade unions, religious organisations and academic institutions.

There has been in recent years considerable debate on both human rights and Panchayati Raj. A discussion on grassroots institutions like Panchayati Raj and their linkages with the question of human rights opens up several new issues. The overall thrust of human rights is to create a social order where not only human dignity is guaranteed but the necessary socio-economic conditions are created for fuller development of every individual and thereby the entire population. This calls for not only freedom and emancipation of the oppressed but creation of institutions (like Panchayati Raj) which are democratic and conducive for such a purpose (Haragopal, 1994). We have undertaken the present study in the context of the line of thinking outlined above. The findings and the suggestions emerging from the study are summarised below. Being a pilot study with time and resource constraints, we have attempted only to identify human rights issues at the grassroots level (Panchayat) and the role of different agencies (including the Panchayat) in HR protection, promotion and education. As a corollary to the main agenda of investigation, the scope of reviving NPs as a participatory justice process is also examined.

As 72 per cent of the sample belong to socially, economically and educationally backward communities, the conclusions of the study can be taken as the opinion, views, and concerns of the backward and marginalised sections of the village population who are most vulnerable to HR violations and are deprived of involvement in different forms of participatory interventions in mainstream socio-economic life.

It is a well-documented fact that women are more exposed to exploitation and HR violations; they constitute 47 per cent of our sample and their views on different issues also carry much significance in a study like the present one. The high rate of unemployment reported by the respondents is also a reflection of the poor attention paid by the society and the panchayat in particular, on the right to development. Since 46 per cent of the respondents belong to the age group of 31-50 years, their views are taken as a balanced one.

Irrespective of the sample categories, a higher percentage of the respondents (83%) has reported general awareness about human rights; however the level of awareness is comparatively low about HR protection, leaving scope for HR interventions. The important HR issues in the study area are unemployment, police torture, and interventions in matrimonial cases, dowry-related atrocities, torture etc. Reports on custodial atrocities and death due to poverty are few; however the existence of dowry-related atrocities have been reported by 50 percent of the respondents. HR violations by the police have left 35 per cent of the respondents dissatisfied with the police machinery in HR interventions and dispute settlement. It may be remembered that 40 per cent of the cases registered with the NHRC in 2001 has been against the police apparatus. The need for reforming the dealings of the police towards the public and HR education among them are the relevant inferences drawn from our study.

It is observed that people in the study area are well aware of HR violations in the locality, though the political leadership (in terms of the opinion of elected members of the Panchayat) is reluctant to accept the existence of HR violations. Our analysis of the violations reported (220) is an eye-opener on the nature and gravity of HR violations in rural Kerala. Our study endorses the dominant view that weaker sections including women are the worst victims of HR violations in work places, households, police station, public places etc. About 15 per cent of the reported violations involve SCs and STs and 14 per cent of the cases are connected with police atrocities. Dowry-related atrocities (7%) also need to be viewed seriously as they involve cases of exploitation of women and a threat to the institution of family. We have observed that the study area provides much scope for HR intervention by NGOs, COs, HROs etc. Majority of the respondents (62%) hold the view that HROs are instrumental in awareness, prevention and education in the study area; the Panchayat is found to be contributing the least in the above matter, compared to other agencies involved in the process, namely HROs, COs and NGOs.

Though Gramasabha is a popular forum under the PRI system, only very few HR issues are raised in it; however the scope of an HR working group in gramasabha is welcomed by 90 per cent of the respondents. Members of COs and SHGs are the strong advocates of the above proposal. There has been general agreement on increasing involvement of panchayats in dispute settlement/HR issues. Similarly the majority also welcomes the potential role of NPs, as a participatory institution in dispute settlement. The Panchayat as a potential forum for HR education is also widely shared but opinion is not in favour of reviving the old form of NPs, as it leaves possibilities of political interference in its operation.

Opinion on institutional preferences in dispute settlement has brought out interesting conclusions with a lot of policy implications. Though multiple preferences are expressed, the most preferred dispute settlement/rights protection agency is HROs and the least preferred one is political party. The respondents approve free legal aid agencies and the judiciary next to HROs. Despite the critical attitude to the police machinery, respondents have much higher preference for it over political parties and the panchayat.

In the light of the present study the following suggestions are made:

1. There is a case for a state-wide research study to document participatory systems of dispute settlement.
2. Though human rights issues are not new to the macro situations, awareness and education need further strengthening at local levels, especially in the clusters of weaker sections and women.
3. As there is a general tendency to turn to local police in cases of HR violations/ disputes, human rights education and training have to be imparted to reform the police service.
4. Since elections to the offices of local self governments are contested on political ideologies, NPs may not be free from political interference especially in a highly politicized state like Kerala. Hence instead of reviving NPs, it would be better to strengthen HROs, and
5. Panchayats, being closer to the people in multifarious ways, can meaningfully undertake HR promotion through educational programmes for their officials/ elected members as well as the public.

There are strong indications in the government circles that participatory interventions in the justice process would be promoted in the programme of decentralised governance in Kerala. The Chief Minister of Kerala has told the delegates of the National Legal Workshop and Seminar at Kollam on 27th December 2001 that the State government was considering a proposal to establish NPs at the grama panchayat level to deal with minor offences. He hoped that NPs would lessen the burden of civil courts where more than 4 lakh cases had been pending for many years (Panchayati Raj Update, ISS 2001).

ANNEX - 1

The Universal Declaration of Human Rights 1998: A Synopsis

This abbreviated version of the 30 Articles of the Universal Declaration of Human Rights provides an overview of the principal rights and freedoms that are every person's birthright.

The first two articles are fundamental principles underlying all human rights. Articles 3 to 21 comprise civil and political rights. Articles 22 to 27 refer to economic, social and cultural rights. The last three articles provide a framework of solidarity safeguarding the universal enjoyment of all human rights.

Article 1	Right to freedom and equality in dignity and rights
Article 2	Freedom from discrimination
Article 3	Right of life, liberty and security of person
Article 4	Freedom from slavery and servitude
Article 5	Freedom from torture or degrading treatment
Article 6	Right to recognition as a person before the law
Article 7	Right to equal consideration before law
Article 8	Right to remedy through a competent tribunal
Article 9	Freedom from arbitrary arrest or exile
Article 10	Right to a fair trial or public hearing
Article 11	Right to be considered innocent until proven guilty
Article 12	Freedom from interference with privacy, including home, family and correspondence
Article 13	Right to freedom of movement and residence in one's own country and to leave and return at will
Article 14	Right to asylum
Article 15	Right to a nationality and freedom to change it
Article 16	Right to marriage and protection of family
Article 17	Right to own property
Article 18	Freedom of belief and religion
Article 19	Freedom of opinion and information
Article 20	Right to peaceful assembly and association
Article 21	Right to participate in government and in free elections and to equal access to public service
Article 22	Right to social security
Article 23	Right to work and fair pay for work
Article 24	Right to rest and leisure
Article 25	Right to adequate standard of living for health and well-being
Article 26	Right to education
Article 27	Right to participate in the cultural life of the community
Article 28	Right to social order assuring human rights
Article 29	Responsibility to community essential to free and full development of the individual
Article 30	Freedom from State or other interference in any of the above rights

ANNEX - 2

Names and Address of NGOs included in the rolls of the Kerala State Human Rights Commission

1. Human Rights Protection and Welfare Council of Kerala, Kadakkavoor P.O., Thiruvananthapuram.
2. Consumer Guidance and Research Society of India, P.O. Aluva, Ernakulam District.
3. Sevanam Centre for Social Services, Aluva P.O., Ernakulam District.
4. Legal, Industrial and Socio Educational Society of India, Kothamangalam P.O., Ernakulam District.
5. Social Welfare and Educational Society, Mudikkal P.O., Ernakulam District.
6. Human Rights Protection Council, Perumbavoor P.O., Ernakulam District.
7. Consumer Guidance and Complaint Cell, Moovattupuzha P.O., Ernakulam District.
8. OISCA-International, C-8 Jawahar Nagar, Nadakkavu, Calicut-673 006.

REFERENCES

A Handbook of Travancore, Kerala Vilasom Press, Trivandrum (1881): Vol. I, p. 327.

A Judicial History of Travancore (1932): Published on the Occasion of the Golden Jubilee of the High Court, Appendix I, High Court, p. iii.

Baxi, Upendra (1980): *The Crisis of the Indian Legal System*, Vikas Publishing House, p. 324.

Baxi, Upendra (1986): "Participatory Justice: A Critique of Nyaya Panchayats in Karnataka", in George Mathew (ed) *Panchayati Raj in Karnataka Today: Its National Dimensions*, Institute of Social Sciences and Concept Publishing Co. New Delhi, pp. 76-87.

Cochin Information (1944): Vol.III, No. 2, Issued by the War Publicity Department, February-March, p.10.

Cochin Information (1945): Vol. IV, No. 4, April-June, p. 20.

Cranston, Maurice (1973): *What are Human Rights?*, The Bodley Head London, p. l.

Government of India, Law Commission (1958): *Reform of Judicial Administration*, Fourteenth Report.

Government of India (1973): *Report of the Expert Committee on Legal Aid: Processual Justice to the People*, Chairman Justice V.R. Krishna Iyer.

Government of India (1977): *Report on National Judicare: Equal Justice-Social Justice*, Chairman, Justice Bhagawathi.

Haragopal, G (1994): *"Panchayati Raj and Human Rights"*, Occasional Papers-6, Renewing Local Self Government in Rural India, Govt. of India, Ministry of Rural Development, Department of Rural Development, NIRD, Hyderabad.

Haragopal, G (1997): *Political Economy of Human Rights: Emerging Dimensions*, Himalaya Publishing House.

Institute of Social Sciences (2001): *Panchayati Raj Update*, Vol. VIII, No.12, 96, December, p. 3, 8.

Institute of Social Sciences (2002): *Panchayati Raj Update*, Vol. IX, No. 9, 105, Sept, p. 5.

Institute of Social Sciences (2002): "Decentralisation as an Instrument of Human Rights Promotion–I" (Edited excerpts from 'Local Rule: Centralization and Human Rights'- published by International Council on Human Rights Policy, Geneva), *Panchayati Raj Update,* Vol. IX, No.10, October, p. 6.

Institute of Social Sciences (2002): "Decentralisation as an Instrument of Human Rights Promotion–II" (Edited excerpts from 'Local Rule: Centralization and Human Rights'- published by International Council on Human Rights Policy, Geneva), *Panchayati Raj Update,* Vol. IX, No.11, 107, November, p. 6-7.

Institute of Social Sciences (2002): "Decentralisation as an Instrument of Human Rights Promotion–III" (Edited excerpts from 'Local Rule: Centralization and Human Rights'- published by International Council on Human Rights Policy, Geneva) *Panchayati Raj Update,* Vol. IX, No.12, 108, December, p. 6.

Innes, C.A (n.d): Malabar (Gazetteer), FD Evans (ed) Vol. I and II Reprinted by the State Editor, Kerala Gazetteers, Thiruvananthapuram, 1997, p. 385.

Iyer, Krishna V.R (1979): *Of Law and Life*, Vikas Publishing House, p. 141.

Jacob, Alice (n.d): *"Rural Litigation: Role of Nyaya Panchayats as Agents of Grassroots Justice"*, Appendix II in S.N. Mathur (1997), *Nyaya Panchayats as Instruments of Justice*, ISS, New Delhi.

Jaswal, Paramjit S (1996): *Directive Principles: Jurisprudence and Socio-economic Justice in India*, APH, Publishing Corporation, New Delhi.

Koenig, Shulamith (1994): "Defending the Future", in Reoch Richard (ed.) *Human Rights: The New Consensus*, Regency Press in Association with the UNHCR, p. 45, 46.

Koshy, Ninan (2000): "Human Rights under Globalisation", *South Asian Journal of Socio Political Studies*, Vol. 1, No. 2, June-December, pp. 14-20.

Kushawaha, R (1977): *Working of Nyaya Panchayats in India: A Case Study of Varanasi District*, Young Asia Publications, New Delhi, p. 105.

Menon, Madhava N.R (2000): Sri Chithira Thirunal Memorial Lecture, University of Kerala, March.

Menon, Madhava N.R (2000 a): "Social and Economic Rights: Poverty is Greatest Threat", in Mathews, George Chunakara (ed) *Human Rights 2000*, Institute of Human Rights, Vigil India Movement, Bangalore, pp. 5-12.

Mathur, S.N (1997): *Nyaya Panchayats as Instruments of Justice*, Institute of Social Sciences, Concept Publishing Company, New Delhi.

Mathew, George (2003): "Panchayati Raj Institutions and Human Rights in India", *Economic and Political Weekly*, January 11, Vol xxxviii, No. 2, pp. 155-162.

Pal, Mahi (1999): "Functioning of the National Human Rights Commission-A Brief Review", in *State of Human Rights in India 1999*, Sponsored Publication, Indian Social Institute, New Delhi, p. 7.

Pal, Mahi (2000): "Panchayats and Human Rights", *Kurukshetra*, Vol. 48, No.11, August, pp. 32-36.

Raghaviah, Maleeha (2002): *Jagratha Samithies Hold the Key*, The Hindu Daily, July, 29.

Rajan, J.B (2002): "Declining Indigenous Regulations on Fishery: A Case of Kadakkodi (Sea Court) in Kerala, South India", Fisheries Research Cell, Programme for Community Organisation (PCO), Thiruvananthapuram (paper presented in the XIIIth

International Congress on Legal Pluralism and Unofficial Law in Social, Economic and Political Development", organised by the Commission on Folk Law and Legal Pluralism, Nepal, Chiang Mai, Thailand, 7-10 April, 2002.

Reoch, Richard (1994): *Human Rights: The New Consensus*, Regency Press in Association with the UNHCR, p. 45, 46.

Singvi, L.M (1977): Foreword to *Working of Nyaya Panchayats in India: A Case Study of Varanasi District*, Kushawaha R, The Institute of Constitutional and Parliamentary Studies, Young Asia Publications, New Delhi, p. ii and iii.

Part - IV

DECENTRALISATION, TRANSPARENCY AND SOCIAL DEVELOPMENT

7

ACCOUNTABILITY AND TRANSPARENCY IN DECENTRALIZATION: THE KERALA SCENARIO

INTRODUCTION

In recent years, there has been considerable debate on the merits of government decentralization. While this discussion has, in the past, focused on the provision of greater variety of public goods that may result from decentralization, more recently, greater emphasis has been placed on the role that decentralization may have in curtailing corruption.[1] Recent empirical evidence [Fisman and Gatti, 2002a], based on cross-country regressions, suggests that decentralization may in fact be effective in reducing corruption. Similarly it is maintained that local democracy is more effective than national democracy, because the citizens' participation in local decision-making continues beyond the ballot box (Stockmayer, Albrecht 1999). Thus the trend towards greater decentralization has been motivated by disenchantment with previous centralized modes of governance, owing partly to a perception that monolithic government bred high levels of rent seeking, corruption and lack of accountability of government officials. An important research question, therefore, concerns the effects of decentralization on corruption. Can decentralization be a useful institutional reform to reduce corruption, or might corruption increase as political power shifts downward [Bardhan and Mookherjee 2005]?.Opinions on this vary widely. According to one view [Prudhomme 1994], corruption would probably be more widespread at the local level than at the national level. This, by the way, might not be bad in terms of redistribution because the benefits of decentralized corruption are probably better distributed than the benefits of centralized corruption. Corruption is costly in terms of allocative efficiency because it leads to a supply of services for which kickbacks are higher rather than to those for which there is a demand. It is also costly in terms of production efficiency because it leads to corruption-avoiding strategies that increase costs, favor ineffective technologies, and waste time. Though corruption is hard to assess and measure[2], several reasons can be given for greater corruption at the local level than at the national level [Prudhomme 1994]:

- More opportunities for corruption in taxation and authorizations probably exist at the local level.
- Local politicians and bureaucrats have more discretion in matters of governance, which in fact are the major theoretical advantages of decentralization, and
- Fewer obstacles to corruption exist at the local level (as the distinction between the bureaucrat and the politician is less rigorous at the local level).

According to Mello, Luiz de and Barenstein, Matias (2002) the empirical evidence underscores the need for caution in the use of decentralization as a tool for improving governance. For decentralization to be a catalyst for improved governance and accountability in government, appropriate economic and political institutions are needed to insulate the decentralization process from excessive capture of the benefits of government provision by local, rather than national elites, and to ensure that sub-national governments operate under hard budget constraints. Fisman and Gatti [2002b] have observed a strong positive relationship between the proportion of a state's expenditures derived from federal transfers, and corruption, as measured by the number of convictions of public employees for abuse of public office. Their results on federal transfers suggest that decentralizing government expenditures may not be beneficial unless accompanied by decentralization of revenue generation.

In a study of political decentralization experience in 16 countries and 7 Indian states (Andhra Pradesh, Kerala, Karnataka, Madhya Pradesh, Maharashtra, Rajasthan and Uttar Pradesh), Keith McLean cites corruption and lack of transparency in decisions as common dangers in the process; information through public access to voting records and local government meetings, vigilance committees to monitor performance of local government with power to suspend works and initiate legal action, monitoring by press, advocacy NGOs, CBOs etc are the key accountability principles/mechanisms (McLean, Keith 2001).

Though the above contentions have to be adequately brought out through elaborate empirical studies across countries having genuine decentralized governance as part of administrative reform, the question of decentralized corruption raises interesting issues for a country like India. The subject of corruption has been an integral part of India's economic, political and social life since achieving independence from Britain in 1947, and probably even before. Political and bureaucratic corruption,at different levels of the government, have been viewed as plaguing the implementation of development programmes and policies (Kohli, Suresh 1975). The advent of local governments with powers to plan for economic development and social justice has added a new dimension to the debate on corruption; how far participatory institutions/processes in local governance can ensure transparency and freedom from corruption? The state of Kerala, well known for its history of social mobilization for better education, health, PDS and for fundamental structural changes such as land reforms, has been a pioneer in implementing decentralization in governance. Signals of achievements and failures from the state, on different aspects of decentralization, would hold valuable lessons for other states as also decentralization around the world. It is against the above background that an attempt is made to critically look at the accountability and transparency mechanisms in the decentralization process of Kerala. The study introduces the issue of corruption in the specific context of decentralization in governance, mainly in terms of theoretical and empirical formulations so as to contextualise the issue with the scenario of Kerala. The theoretical perspectives on the issue of corruption are also reviewed. An overview of the strategies and safeguards evolved/instituted to ensure transparency and address corruption in the decentralized governance system of Kerala is attempted. Concluding observations on the Kerala experience in the task of corruption-free and transparent local governance are presented at the end.

CORRUPTION IN GOVERNANCE: THEORETICAL PERSPECTIVES

Since accountability and transparency in governance are issues directly related to corruption, a brief introduction to corruption in governance would be meaningful. The agenda of corruption and moral turpitude is being discussed since Plato or even before him. Dating back the fourth century B.C. in India *Vishnugupta (Kautilya)* stated "forty" ways of embezzlement by the government officials. *Kautilya* found out the root of corruption in human psychology wherein he compared human tendency with that of a horse, who appears to be noble and silent while pegged to a place but starts showing its true nature when attached to a chariot. So is the case with persons who, when not in job, appears to be honest, sincere and noble but once assured of a permanent employment and given reward, start revealing their true color by assuming innumerable forms to abuse the power of authority (Gupta, I.D 2002).

Until the 1980's, scholarly research on corruption was largely confined to the fields of sociology, political science, history, public administration and criminal law. Since then, economists have also turned their interest to this topic largely on account of its increasingly evident link to economic performance. Since the early 1990's, there has been a virtual explosion of academic writing on the economics of corruption (Abed and Gupta 2002). A considerable part of this research activity took place at the IMF and the World Bank. The latter has recently adopted a framework for addressing corruption as a development issue, in the context of assistance it provides to member countries. The World Bank framework has the following components:

- Preventing fraud and corruption in Bank-financed projects by improving procurement and financial management policies related to the implementation of projects
- Helping countries in their efforts to eliminate corruption
- Taking corruption more explicitly into account in the formulation of country assistance strategies, and

- Adding voice and support to the international efforts to reduce corruption, including collaboration with regional development institutions on anti-corruption measures.

Other players on the international scene (like NGOs, Civil Society Organizations, media etc.) also became interested in more transparent and accountable conduct of economic management. In December 1996, the United Nations General Assembly adopted a Declaration Against Corruption and Bribery in International Commercial Transactions, as recommended by the UN Economic and Social Council. The role of the IMF in corruption-free governance is reflected in its core activities: surveillance of economic policies of member countries, financial support for adjustment programmes and technical assistance to strengthen economic and financial management.

Scholars on corruption have traditionally tended to deal with the definitional issue at best as an unavoidable irritant (Pavarala, Vinod 1996). We are indebted to Heidenheimer et al (1989) for organizing the maze of definitions of corruption available in the social science literature around three basic models: 'public office-centred', 'market-centred', and 'public-interest centred'. 'Public office-centred' corruption includes such behaviour as bribery, nepotism and misappropriation (Nye, Joseph S 1967). Van Klaveren conceives 'market-centred' corruption in terms of a civil servant who regards his public office as a business, the income of which he will, in the extreme case, seek to maximize. The office then becomes a 'maximizing unit'? (cited in Heidenheimer et.al, 1989). The 'public-interest centred' definition views corruption as existing whenever a power holder who is charged with doing certain things, i.e., who is a responsible functionary or office holder is by monetary or other rewards not legally provided for, induced to take actions which favour whoever provides the rewards and thereby does damage to the public and its interests (Friedrich, 1966).

According to the UN Human Development Report (1998), the diverse manifestation of corrupt transactions have been identified as ranging from-acceptance of money or other rewards for awarding contracts, violations of procedures to advance personal interests, including kickbacks from development programs or Multinational Corporations, pay-offs for legislative support, and the diversion of public resources for private use, to overlooking illegal activities or intervening in justice process. Forms of corruption also include nepotism, common theft, overpricing, establishing non-existent projects, pay roll padding, tax collection and tax assessment frauds.

At this point it would also be useful to distinguish between 'grand corruption' on the one hand and routine 'petty corruption' on the other. Grand corruption has been defined as 'the misuse of public power by heads of state, ministers and senior officials for private pecuniary gain' (Moody Stuart, 1994). Petty corruption means the solicitation or extortion of small payments by low-level officials in order to expedite business by cutting through red tape, or to do what they are supposed to do any way. Petty corruption, in this form, is often referred to as 'speed' or 'grease' money. It is no less significant precisely because it bears most heavily on the poorest and the weakest sections of society who can least afford to make payments for services to which they are any way entitled (Doig, Alan and Theobald, Robin 2000). Heidenheimer et al (1989) categorise corrupt behaviour into 'petty corruption', 'routine corruption' (including acceptance of gifts by public officials, nepotism, and graft) and 'aggravated corruption' (including kickbacks and organized crime). Though not comprehensive, Transparency International regularly publishes Corruption Perception Index (CPI) under which a number of countries are ranked according to perceptions of corruption registered in a range of relevant surveys.[3]

IMPACT ON THE ECONOMY AND SOCIETY

Apart from undermining confidence in the government, and its moral authority, corruption has very real economic costs.[4]

- It is likely to make the misallocation of resources worse, because corrupt officials are unlikely to press for change in the regulations from which they get rich. It may also aggravate income inequalities and poverty.
- Empirical evidence shows that corruption negatively affects economic growth. Since corruption acts as a tax on enterprise, it raises costs and reduces incentives to invest.
- Often corruption also tends to shift government spending away from socially beneficial investments [such as health, education, good roads and communication] toward unneeded 'white elephant' projects or lower – quality infrastructure.
- Corruption also reduces domestic savings and investment and stimulates capital flight [Wolf, Thomas and Gurgen, Emine 2000].

Evidence presented in an important paper supports the argument that corruption can reduce growth by:

- Increasing public investment while reducing its productivity
- Increasing public investment that is not adequately supported by non–wage expenditure on operation and maintenance. Evidence also shows that higher corruption is associated with higher expenditure on wages and salaries
- Reducing the quality of the existing infrastructure, and
- Decreasing the government revenue needed to finance productive spending [Tanzi, Vito and Davoodi, Hamid 1998].

Econometric studies on the effects of corruption have revealed that it reduces:

- Investments and rate of growth [Mauro 1995], where growth is a function of investment]
- Expenditure on education and health [Mauro 1997 a], which does not lend itself easily to corrupt practices on the part of those who make budgetary decisions, and
- Expenditure for operation/maintenance, the productivity of public investment and tax revenue [Tanzi and Davoodi 1997].

Thus corruption has been around for a very long time and will be around in the future unless governments can figure out effectively ways to combat it (Mauro, Paolo 1997 b). Recent decades have witnessed increasing corruption in the political and economic life of many of the top world leaders and even the economic system of communism is stated to have fallen prey more to corruption than to its inner or inbuilt ideological weaknesses (Gupta, op cit 2002). National governments and international agencies like IMF/IBRD have increasingly become aware of the dangers of corruption and the need for combating the menace through legislative, political, economic, socio-cultural and administrative measures. Perceptions/constructions of social scientists and other elite groups (for example bureaucrats, industrialists, judges, journalists and politicians) may differ as to the social, cultural and political consequences of corruption. But in the ultimate analysis it is found that corruption could be more effectively contained through strengthening the institutions of democracy itself. It is only properly functioning democracy, which could lead the country towards freedom from corruption (Gupta, op cit 2002). The fight against corruption is, thus, intimately linked with the reform of the state (Vito, Tanzi op cit 1998); and decentralization is one of the reforms worth pursuing.

INDIA: REFLECTIONS ON CORRUPTION

In a special survey of Indian economy, The Economist [1991] suggested that corruption is responsible to a substantial extent for the country's economic problems. The elites' perspective of the economic consequence of corruption is that large scale corruption results in the distortion of development goals and diversion of the fruits of development away from the intended beneficiaries [Pavarala, op cit 1996].

The rating of corruption as a social problem by a group of elites is summarized in Table 1. The differential ranking of corruption as a social problem depends to a large extent on the configuration of power [Pavarala, op. cit 1996]. The Andhra Pradesh elites interviewed in the above study offered three contending doctrines regarding the consequences of corruption in India, namely, the doctrine of minimal consequences, horrible consequences and positive consequences. [Pavarala, op cit 1996]. The differential construction of the consequences of corruption is indicated by the differential ranking given to corruption as a social problem in India by the various elite groups (See Table 1). Businessmen and bureaucratic elites who put forward the doctrine of positive consequences contents that corruption, rather than being harmful, has some beneficial consequences; one popular proposition is that corruption helps grease the creaky administrative machinery and provides much needed speed to bureaucratic decisions (Pavarala op.cit 1996). It may be observed that the functionalist discourse on corruption, which was particularly prominent in the 1960's contents that corruption has some positive consequences for the developing countries (Leys, 1989).

Corruption originates not in any system but is an integral part of the mindset of the man himself. Of late, however, it has become a cause of concern owing to deeper proportions and penetrations it has taken in the system, infringing upon and distorting the institution and forces of growth and development, in many of the less developed countries where growth is the basic priority, to remove poverty absolute and relative, and establish social and economic justice (Gupta, op.cit 1996).

Table 1
Rating of Corruption as a Social Problem by Elites

Elite Group	*Not a Major Problem*	*One of the Major Problems*	*Most Important Problem*
Bureaucrats [N = 15]	9; Reasons:Other more important problems Eg; communalism, religious conflict, casteism.	3; Other major problems cited: poverty, over population, illiteracy.	3; Reasons:'corrodes our culture'; 'retards progress'.
Industrialists [N = 10]	7; Reasons:Corruption as positive effects; other more important problems, eg., low productivity, unemp-loyment, inefficiency.	2; Other major problems cited: poverty, over population, under employment.	1; Reasons:'makes country uncompetitive in the global market'.
Judges [N = 12]	0	4; Other major problems cited: Communalism, regionalism, separatist movements, overpopulation.	8; Reasons: 'destroys civilized life'; 'threatens democracy'; threatens value system'.
Journalists [N = 12]	0	3; Other major problems cited: poverty, inflation, communalism.	9; Reasons: disaster for democracy'; 'threatens our way of life'; 'subversive rule of law'; 'evil facing the nation'.
Politicians [N = 11]	2; Reasons:Other more important problems e.g, communal violence, poverty.	3; Other major problems cited: communalism, underemployment, separatist movements.	6; Reasons: 'threatens our very existence'; 'marginalizes development'; distorts welfare goals'.
N = 60	18	15	27

Source: Pavarala, Vinod [1996], Interpreting Corruption: Elite Perspectives in India, p. 72.

A study by Huther and Shah has placed India under countries having "fair governance based on Quality of Governance Ranking". According to them, decentralization as an anti-corruption instrument may improve accountability and may increase sense of social purpose for public officials; its effect may be 'medium' in weak, and 'low' in fair and good governance environments [Huther and Shah 1998].

REVIEW OF CORRUPTION UNDER DECENTRALIZATION

A brief review of the evidence from cross-country regressions concerning the relation between decentralization and measures of corruption and governance, undertaken by Bardhan and Mookherjee [2005], would be relevant in this context. The studies reviewed by the authors evaluated:

- The relation between a measure of expenditure decentralization and per capita delivery of different forms of infrastructure
- The effects of expenditure decentralization and revenue decentralization on corruption, and
- The relation between appointment decentralization and corruption.

Bardhan and Mookherjee observe that cross- country empirical studies fail to provide robust evidence of benign effects of decentralization on governance. According to them, within-country empirical studies and evidence in other countries concerning determinants of accountability and capture of local governments indicate that the effects of decentralization on corruption and government accountability are complex and cannot be summarized by simple, unconditional statements.

However, with decentralization and community involvement, it is hoped that intermediaries and private contracting would be reduced/watched closely so that leakages in development expenditure could be minimized. But how far such safeguards (inbuilt and incorporated over time) have served the process is a matter to be looked into. Recent reports on corruption under decentralization in India are illustrated in Chart 1.

Chart 1
India- Recent Reports [2004-05] on Corruption under Decentralization

State	Report on Corruption
Bihar	A total of about 200 Mokhiyas have been dismissed so far on various charges since Panchayati Raj was restored in the state after elections in April 2001.
Goa	The Socorro civic & consumer forum exposed panchayat scams involving contracts for garbage collection.
Karnataka [5]	The department of Panchayati Raj & Rural Development has detected 90 cases of financial & other irregularities by senior officials working in the engineering division in various Taluk & District Panchayats.The C & AG report for the year ending March 31 2002 indicated misappropriation of stores worth Rs. 1.18 crore in just 3 Zilla Panchayat engineering divisions.
Kerala	A local fund audit has pointed out irregularities to the tune of Rs. 7 lakh in the execution of public workers under the PPC in Mangalapuram, Chemmaruthy, Karode& Kattakada
Orissa	Officials of the vigilance department raided the Pattamundai Block Panchayat office in Kendrapara district on August 13 2004 & seized many incriminating documents, fake seals & work orders
Punjab	Surpanch of Ajit Gill village was suspended for misuse of Panchayat funds to the tune of Rs. 10.5 lakh.
Rajasthan	A public hearing organized on July 21 2004 in Bikaner district revealed that about Rs. 30 lakh has been swindled from the Panchayat fund during the past 5 years.

Source: Institute of Social Sciences, 'Panchayati Raj Update', various issues, 2004 & 2005.

ACCOUNTABILITY AND TRANSPARENCY IN DECENTRALIZATION: CRITICAL LESSONS FROM KERALA

Kerala has been a pioneer in initiating a number of progressive social, cultural, educational and economic movements and policies over the last few decades. The latest in the series of mass mobilization for socio-economic change is the People's Campaign for decentralized participatory planning launched in 1996. The primary objective of the campaign has been to empower the local bodies to prepare and implement their own development plans. Kerala is one of the few states, which have actually devolved almost all the 29 subjects in the XIth Schedule of the Constitution to the local bodies. Adequate funds and required staff are placed at the disposal of the local bodies to facilitate their effective functioning. Being a highly literate society with a long tradition of mass mobilization, institution building, political and social experiments, Kerala has also witnessed proliferation and interface of new participatory institutions which have stepped up their scale of involvement in the process of grassroots planning for economic development and social justice (Sankaran and Vijayakumar 2001).

One of the outstanding merits, visualized for participatory planning is that there would be greater scope for ensuring transparency, accountability and minimizing corruption in decision making, selection of beneficiaries, execution of public works etc. in all types of local governments. In the decentralization experiment of Kerala, the strategy evolved to ensure accountability, transparency and corruption-free governance has been a set of formal and semi-formal mechanisms (see Chart 2). The former consists of Local Fund Audit, Performance Audit, Grievance Redressal Committees and Ombudsman. Public access to information, Beneficiary Committee, criteria for beneficiary selection, Social Audit and Citizens' Charter come under the latter. In the present study, selected transparency and corruption-control mechanisms are discussed elaborately, namely Performance Audit, Beneficiary Committee, Beneficiary Selection, Social Audit and Citizens' Charter.

Since substantial funds have been given to the local governments (35-40 per cent of total plan outlay), accountability system acquires special importance. Traditional corruption-control and accountability mechanisms like official audit and inspection, can help decentralization to a considerable extent. Kerala has attempted two innovations–first, setting up Performance Auditing teams by pooling surplus staff in the Panchayats, Municipalities, Rural Development Department and the Secretariat to conduct regular auxiliary audit with a view to correcting mistakes as and when they occur and guiding local governments in maintaining the proper system: second, setting up a Technical Audit Team consisting of senior engineers to investigate complaints of malfeasance in public works. Regular IEC campaigns, participatory structures right from planning upto monitoring, transparency provisions and scheme formulation framework and spelling out the due process in various kinds of decision making, which fix the rational boundaries to autonomy, are the important examples (Vijayanand, 2001). In addition to the Ombudsman, new accountability mechanisms have also come up like the Citizens'

Chart 2
Decentralized Governance in Kerala: Accountability, Transparency and Anti-corruption Mechanisms

Formal	Semi-Formal
➢ Local Fund Audit ➢ Performance Audit ➢ Grievance Redressal Committee* ➢ Ombudsman**	➢ Public access to Information*** ➢ Beneficiary Committee ➢ Beneficiary Selection ➢ Social Audit ➢ Citizens' Charter

Source: Developed from, Rural Decentralization in India, World Bank, September 2000, pp. 111-112.

* Consists of honest people who can order an investigation on complaints of technical irregularity.

** Added by the author for the present study.

*** It has been decided following the recommendations of S.B. Sen Committee that all plan documents including those relating to beneficiary selection, bills and vouchers or works are public documents which any citizen can access.

Charter. The accountability, transparency and anti-corruption mechanisms are introduced below.

PERFORMANCE AUDIT

Performance audit is a non-traditional approach to audit (the traditional type of audit of local bodies is done by the Accountant General, Local Funds Dept. and Finance Dept-Inspection Wing). It promotes the right to information involving concurrent evaluation of projects. Rather than a mechanism to bring to book persons responsible for lapses, it aims at improving project performance/ transparency through frequent interventions. Performance audit also involves spot verification, inspection and interaction with project beneficiaries. Its operational frame consists of evaluation of assets/services in terms of quality, sustainability, time- frame of execution, access of the people/peoples' organizations, to the planning process etc (Government of Kerala, ICDD 2000). The advantages of the process are:

- Quarterly periodicity of the exercise and notification by the regional/district auditing officer
- Opportunity available to the public to bring to official notice complaints, if any
- Circulation of questionnaire by auditing authority to the Panchayat to collect information

- Field visit for spot evaluation of projects and collection of opinion/complaints from the people
- Reporting of audit observations within a week and discussion thereof in the meeting of the Panchayat/Grama Sabha on the questions raised and the reply/clarifications submitted, and
- Interaction with audit officers on issues found unacceptable to the local body for mutually agreeable resolution.

Though performance audit is a welcome step in the interest of transparency and the right to information, it is not clear whether performance audit will prevail ultimately, if there is a conflict with the other systems of audit.

BENEFICIARY COMMITTEES (BCs)

Preferably all the development works are to be executed by a BC. The perceived logic behind BCs is to prevent leakages resulting from the intervening role of the contractors in public works. When a development work is to be undertaken, persons and households who will be benefited by the work are called for a grama sabha meeting and they will be encouraged to form a committee of 10-15 people, including women and scheduled castes, to implement the work. The BC is required to execute a bond with the grama panchayat, and open a bank account of its own. The Committee will be given an advance of Rs. 1 lakh or 25 per cent of the cost (whichever is less). After examining the progress by the expert committee, the second installment is given. The main idea of formation of BC is that the works will be completed in time, cost of the projects will come down, beneficiary contribution will go up and transparency will be achieved. When the BCs actually started functioning they faced numerous difficulties. The role of the government engineers continued to be conspicuous in roles such as supervision, measurement, verification, preparation of bill and the final payment. There are numerous hurdles that these BCs face, given their lack of wherewithal, technical know-how and skilled workers. In many places, *binami* contractors have appeared and the BCs merely served as facilitating mechanisms of the contractors for faster payments. Genuine BCs, which are very few, have never taken a second work if the experience of Kottayam district is any indication (John and Chathukulam, 2001). However, quality of work done by the contractors has improved to some extent because of quick payments. A study conducted by the State Planning Board in 503 best BCs drawn from different panchayats of Kerala suggests that BCs are functioning quite well (Govt. of Kerala, 1999). Field study conducted by John and Chathukulam in Chempu Panchayat, Ernakulam district, Kerala, also endorse the above view.

Though it has been generally claimed that the grama sabhas and BCs have resulted in considerable accountability, transparency and control of corruption, a World Bank study [2000] found that the institutions/mechanisms are not very effective/widespread. In several cases, contractors have replaced the BCs through the backdoor. The public is interested in the early completion of the works rather than their actual cost. Because of increasing development fatigue, local interest and participation in grama sabha are also on the decline.

DFID conducted a study on the social issues in community contracting covering India, Ghana, Bangladesh and Zambia. The Socio Economic Unit Foundation Kerala, which coordinated the study on India, addressed the social issues of construction works under the Peoples' Plan of Kerala. In about 50 per cent of the public works covered by the study during 1999-2000, despite the formation of BCs, contracting system was found to be widespread. The study also noted lack of transparency in project selection, procedure for according the contract and other administrative procedures (SEUF 2001). One of the biggest failure of decentralized planning and implementation has been in the area of community contracting. As genuine BCs could not do the work, contractors masquerading as conveners of BCs took up work in a large number of places. This opened up opportunities for abuse of systems through patronage and deals with elected local governments. Hence Modernizing Government Programme (MGP), Government of Kerala, 2003 has proposed the design of a new Public Works Manual for LSGs that allows community contracting and that facilitates social audit.

BENEFICIARY SELECTION

Decentralization aims at transparency in the process of beneficiary selection as well. To select beneficiaries, (from among those BPL) for development programmes of local bodies, each Grama Panchayat will develop an application form and decide eligibility and priority criteria. After inviting application from the people, a committee consisting of the President, the Ward Member of the locality and others in the village will examine each of the application forms and give marks to them. The applicants with maximum eligibility in the locality will be selected. The Ward Member will read out the select list in the grama sabha. In addition, the application forms along with marks and criteria adopted to select beneficiaries will be printed and displayed in the office of Grama Panchayat (GP) and Anganwadi, for which also funds are made available to GP. Until the printing is complete, a xeroxed copy is to be displayed. Any one can demand access to these documents.

The World Bank study team observes that they have not seen the printed list of the beneficiaries, though a xeroxed list was made available in one GP. Hence the study concludes that:

- The elected leaders do not seem to be convinced about the superiority of the method
- they maintain a xerox list only to comply with government rules
- The method could be abused as they have discretion for deletion / inclusion of existing or new criteria
- The people do not seem to be happy with the method
- There are complaints that better-off people are selected while the worse-off applicants are left out
- People expecting to get higher marks attend the committees and fail to turn up in the grama sabha, and

- The illiterate feel that the method is too technical. To sum up, selection of beneficiaries in the grama sabha meeting is, however, superior, as it can take care of all the above factors and remain free from complexity (World Bank, 2000).

SOCIAL AUDIT

Social audit is a participatory mechanism devised by local bodies in their own way; there is no uniformity in the methodology. Hence we propose to look at social audit with reference to a case study of Nellanad Gram Panchayat (Government of Kerala, ICDD, 2000). As a prelude to local audit, it was decided to undertake all the public works through Beneficiary Committees. The first audit was carried out in an open assembly consisting of 20 and 60 elected members respectively from the ward level NHG and Gram Sabha, members of panchayat level NHG, health promotion *samiti*, implementing officers and elected representatives. The project components subjected to social audit were:

1. Quality, fund utilization and implementation pattern of construction activities
2. Working of beneficiary committees
3. Time-frame of implementation
4. People's mobilization and additional resource mobilization
5. Promotion of people's right to information
6. Effectiveness of NHGs
7. Interaction between people, elected representatives and officials, and
8. Functioning of Voluntary Technical Corps (VTC) having advisory powers on technical issues.

The methodology adopted in Nellanad had the following steps:

- Formation of an internal audit *samithi* as recommended by the Gram Sabha, consisting of technical experts, professional auditors, people of high social standing/acceptability, social activists etc
- Collection of information, documents, reports etc. from Panchayat bodies, implementing officers, beneficiary committees and related institutions by the *samithi*
- Preparation of an audit report covering general administration, development projects, services, operation of popular committees etc
- Circulation of the report to NHGs for comments
- Consideration of the report by the Panchayat, Gram Sabha and Development Seminar, and
- Finalisation of the report and onward transmission for performance audit and local fund audit.

The social audit report covered individual beneficiary schemes, community assets, public institutions and social service organizations. As in the case of performance audit, no system has been devised for the resolution of possible conflict that may arise from social and other systems of audit. If social audit is subjected to official modification, its participatory color is lost. MGP of Government of Kerala 2003 proposed the following methodology for social audit (Design and Implementation) as part of strengthening accountability of LSGs:

- To study the various existing audit practices and the scope
- Study the practice of social audit in India and outside from relevant literature
- Discuss the draft policy with stakeholders and key practitioners within and beyond
- Develop draft guidelines
- Conduct action- research in 10-15 local governments
- Hold workshops of experience
- Develop amendments to the Act and Rules and Manuals for social audit
- Train local governments and representatives of the civil society in social audit
- Launch a publicity campaign (March 2006), and
- Identify a set of independent agencies for monitoring (April 2006).

KERALA PANCHAYATI RAJ (PREPARATION OF CITIZENS' CHARTER) RULES, 2004

Rural local governments in the state were required to prepare and publish The Citizens' Charter by November 2004, following the above rules that came into force on May 17 2004.[6] In the context of local governance, a Citizens' Charter is a statement of services, paid and unpaid, that a local government offers to the Citizens for improving their living conditions. It includes the issuance of permits, licenses and certificates for different purposes. A Charter usually enlists the particulars of services, conditions for obtaining those services and the time limit within which the services must be made available to a citizen. But the Charter needs to be backed by an effective reporting[7] and information systems as tools for social and performance auditing. An attempt made in two GPs [Karakulam & Nellanad] in Thiruvananthapuram District towards developing Citizens' Charter as a tool for social audit is indicated in Chart 3.

Apart from the above mechanisms evolved in the wake of decentralized governance in Kerala, for ensuring accountability and transparency and reducing corruption, few other agencies have also been working indirectly towards the issues under discussion.[8]

CONCLUDING OBSERVATIONS

Corruption in governance and delivery of public services in developing countries have been attributed to over-centralization and lack of monitoring mechanisms over bureaucratic behavior. Hence devolution of powers to elected local governments, which are closer to the people, is expected to improve governance and service delivery. However, if political accountability of local governments is limited, decentralization is also prone to leakages and targeting failures [Bardhan and Mookherjee 2000]. Based on an analytical framework to evaluate the effects of decentralization, the authors argue that the mode of financing local governments is a key parameter. Given the limited accountability of local governments, expenditure decentralization accompanied by full devolution of revenue raising authority to local governments may lead to efficiency and equity losses relative to centralization.

Chart 3
Procedure for Developing Citizens' Charter as aTool for Social Audit: Karakulam and Nellanad GPs, Thiruvananthapuram District

• Generation of Idea	Situation analysis by *Grameena Patana Kendram*, under a project 'Responsible Administration: A Management Development Mecharism'.
• Entry activities & Creation of Environment with the involvement of subject experts, elected members officials & project team	Citizens' Charter & Public Information Systems in GP/transferred institutions-Service Boards [explaining institutional services], Attendance Boards [exhibiting attendance of public officials], InformationBoard [ensuring accessibility of information] & Complaint/Suggestion Box [for public use].
• Social Audit	Participatory platforms like Grama Sabha, Kudumbasree, SHGs etc.
• Public Awareness	Citizens' education campaign [to bring Citizens' Charter into public space] through Official & participatory systems [in terms of Training & discussion, animators, Kudumbasree, Clubs, Trade unions, Residents Associations etc].

Source: Developed from Sreekumar and Gopakumar [2005].

The extent of corruption within local governments is also related to availability of information to citizens concerning their entitlements, to the way they are monitored by higher-level governments and media. There is growing evidence that a number of institutional safeguards are effective in limiting capture of local governments by elites and non-minority groups: reservation of positions in local governments for minorities, information media campaigns, mandated village meetings, civic participation mechanisms, and monitoring by upper level governments ....For policy makers, there are some important lessons: decentralization by itself is unlikely to be a panacea for problems of accountability [Bardhan and Mookherjee 2005].

The seeds of corruption, thus, flow along with transfer of finance, functions and powers down to the local governments; to a great extent the success of the participatory approach of grass root planning, monitoring and evaluation depends on the strengthening of people's institutions/forums like Grama sabha, NHGs, Social Audit, BCs, etc. All of them can exercise powers of checks and balances, and act as watch–dogs in the different facets of decentralized governance. Depoliticisation of village elections/governments and a move towards issue–based contests, will go a long way towards the realization of the true objectives of Grama

Swaraj. Lofty principles of transparency, accountability, corruption-free governance etc. of course, hold an appeal for the people at large, but unless they rally round issue-based contexts, many of the ongoing programmes for capacity building in people's institutions may not succeed. In fact one may think of developing a comprehensive index for rating local governments on qualitative principles of transparency, accountability, corruption-free governance etc. along with other quantitative physical measures of performance.

Though several studies have examined various aspects of the process of decentralization in Kerala, the performance of local governments has not been evaluated adequately in terms of transparency, accountability and corruption–free administration. However, mention may be made of a recent attempt to assess the People's Plan Campaign, in terms of corruption control mechanisms. According to the study, the political will of the Left Democratic Front government, contributions of socially committed voluntary organizations and the high level of social development in Kerala were instrumental in controlling corruption at the grassroots (Mohanan, 2000). The limited empirical evidence brought out by the studies that are covered in the present paper gives a mixed lesson; official studies and shared experiences of social activists/elected representatives of local governments maintain that corruption has been effectively dealt with, in the utilization of funds, beneficiary selection etc, and that the principles of transparency and accountability have been kept high in the deliberations/activities of local governments. It is also held that the quality of physical output has improved and that projects are completed without the usually observed delay. However, the studies by the World Bank, DFID and the independent academic exercise referred to above cast a shadow on the official version on management of corruption, transparency and accountability. A number of local governments were found to be going more by the technical import of the stipulations rather than by the 'spirit' of legislations and guidelines. In a number of BCs from which information was sought by the above studies, it was observed that *binami* contracting system was widespread for one reason or other. The existence of two agencies, namely the BCs and the clandestine contractor, may have enhanced the scale of corruption causing greater leakage in the utilization of funds of the above local governments. Given the gaps in our detailed understanding of the efficacy of transparency and accountability mechanisms built into the decentralization exercise, one will have to agree with the broad indications thrown up by media reports: the programme of decentralization of power in Kerala has fallen on bad days after a sunny start[9] (See Chart-2). The eclipse of the grama sabhas could see the resurgence of the politician–contractor–bureaucracy nexus (Krishna Kumar, 2005). It is also noted that the endemic leakage of public resources through artificial escalation of project estimates, manipulation of tenders, use of inadequate and poor quality materials and so on continues despite the great effort made to contain these corrupt practices through the beneficiary committee system and social audit practices (Oommen, 2004).

As already stated, despite their obvious theoretical advantages, performance and social audit mechanisms may not prevail over other parallel systems of audit:

the former is participatory while the latter is official. In the present environment of ambiguity, the latter tends to dominate the former. In all the spheres of local governance, informed participatory interventions must prevail over official interventions. In order to address the issues of transparency, accountability and corruption–free governance in decentralization:

- Objectivity and sustainability of key participatory institutions/interventions [like Grama Sabha, Social Audit, Performance Audit, BCs, Citizens' Charter etc] have to be ensured
- Development of vibrant LSG institutions through [good] governance- linked incentive grants regime [and disincentive for lapses] needs serious consideration, and
- In addition to the development of a comprehensive general index for rating local governments, a Corruption Perception Index [CPI] would be helpful at national and regional levels.

NOTES

[Earlier versions of the paper were presented in the—seminar organized by SV University, Thirupati, AP [later published in 'Dimensions of Crime and Corruption in India' 2005, [ed.] A. Renga Reddy, Serials Publications, New Delhi], International Conference on 'A Decade of Decentralization in Kerala: Issues, Options and Lessons' Trivandrum 7-9 October 2005, organized by ISS, New Delhi and NIRD Foundation Day Seminar, 9-10, November 2005].

1. Ahlin defines corruption in an occupational choice model as extra fees that must be paid by some entrepreneurs. Corruption decreases wages and total output. Two types of decentralization, regional and bureaucratic, are distinguished. Under imperfect mobility, bureaucratic decentralization increases, and regional decentralization decreases corruption. Decentralization also determines the effectiveness of two anti-corruption mechanisms, electoral accountability and efficiency wages. Bureaucratic decentralization hinders both, rendering them arbitrarily ineffective as it increases; by contrast, regional decentralization makes them more effective [Ahlin, Christian 2004].
2. Should corruption be measured by total bribes or some per capita measure? Should corruption be measured by bribes alone, or should it also include costly efforts made by citizens to influence the design or application of laws in their own self interest?. For a brief discussion, see Bardhan and Mookherjee 2005.
3. Corruption Perception Index (CPI) scores relate to perceptions of the degree of corruption based on surveys of business people, academics and risk analysts, and range between 10 (highly clean) and 0 (highly corrupt). Most of the developing countries in Asia have CPI scores at or below 4 and thus have a long way to go to reach a corruption-controlled stage. The CPI score for India, according to the Transparency International measure was 2.8 for the year 2003 and in terms of country-wise ranking, India ranks 71 out of 102 countries for the year 2002 (See TI at www.transparency.org).
4. For a detailed account of corruption in Asia, an analytical framework for improving the effectiveness of anti-corruption policies and programmes and anti-corruption

initiatives in the Philippines, Korea, Thailand and Indonesia, see Bhargava and Bolongaita, 2004.

5. In the midst of increasing corruption enveloping decentralized governance, we have interesting gender perceptions also on the issue. In a symposium on "women representatives of grama Panchayats and the media" organized by Rural Environment Awareness and Community Help (REACH) in association with the Federation of Voluntary Organization for Rural Development in Karnataka and the Hunger Project in Davangere city, Karnataka on May 6 2004, the nexus between male Panchayat representatives and corruption in Panchayats was highlighted by women representatives (ISS, Panchyati Raj Update, May 2004).
6. Urban local governments have already prepared their Citizens' Charter in accordance with the provisions of the rules framed in 2000. See, Rajasekharan, K (2004) for a detailed account of Citizens' Charter.
7. Citizen report cards system, developed by the Public Affairs Center, Bangalore has received international attention. The report cards are an aggregate of public ratings on different aspects of service quality, built on specific random surveys on users of different public services (i.e. utilities in a city). The report cards also include information on corruption encountered.
8. They are Voluntary Technical Corps (VTC), Neighborhood Groups (NHGs), Self Help Groups (SHGs) and Appellate Tribunal. The main function of VTC, an informal body of retired men/women with expertise, is technical scrutiny of projects of local governments and helping them towards formulation of financially and technically viable projects. Intervention of these agencies have also been instrumental in reducing the scope for corruption through improvement in the techno-economic quality of projects, selection of beneficiaries, monitoring, strengthening grama sabha, social audit, beneficiary committee etc. The Tribunal would serve as an accountability-enforcing mechanism against illegal and biased decisions in the regulatory sphere of local governments.
9. See Chart 4.

Chart 4
Overview of Decentralization Laws and their Current Status

Laws/Rules	*Current Status*
Right to Information	Law remains unattested
Citizens' Charter	Not implemented by a majority of local bodies
Social Audit	Never explained to people, Dwindling participation in the grama sabha, Remains a dream
Performance Audit	Status reduced to mere audit of accounts
Grama Sabha	Dwindling participation
Ombudsman	Seven member body reduced to one
Appellate Tribunal	Constituted-fate unclear

Source: Developed from Krishna Kumar, R (2005).

REFERENCES

Abed, George T and Gupta, Sanjeev (2002): "The Economics of Corruption: An Overview", in *Governance, Corruption and Economic Performance*, [eds.] George T. Abed and Sanjeev Gupta, IMF, Washington D.C. pp. 1-16.

Ahlin, Christian (2004): "Corruption and Anti-corruption Mechanisms under Decentralization", *JEL* No. D. 73, h 77, O17,1.

Bardhan, Pranab and Mookherjee, Dilip (2002): Corruption and Decentralization of Infrastructure Delivery in Developing Countries, Forming Part 2 of their *Working Paper* " Expenditure Decentralization and Delivery of Public Services in Developing Countries".

Bardhan, Pranab and Mookherjee, Dilip (2005): 'Decentralization, Corruption and Government Accountability: *An Overview* 'For *'Handbook of Economic Corruption'* [ed] Susan Rose-Ackerman, Edward Elgar.

Bhargava,Vinay and Bolongaita, Emil (2004): *Challenging Corruption in Asia: Case Studies and a Framework for Action*, The World Bank, Directions in Development, Washington D.C.

Doig, Alan and Theobald, Robin (2000): Why Corruption?, in *Corruption and Democratization*, [eds] Alan and Robin, Frank Cass, London, p. 5.

Fisman, R and Gatti, R [2002 a]: "Decentralization and Corruption Across Countries", *Journal of Public Economics*, 83, 325-345.

Fisman, R and Gatti, R [2002 b]: "Decentralization and Corruption: Evidence From US Federal Transfer Programs", *Public Choices*, 113, 25-35.

Friedrich, Carl J (1966): "Political Pathology", *Political Quarterly*, 37: p. 74.

Government of Kerala (1998): Report of the Committee on Decentralization of Powers (S.B. Sen, Chairman).

———, State Planning Board (1999): *Gunabhokthrusamithy Pravarthana Margarekha* (Malayalam).

———, (2000): *"People Friendliness- Transparency: New Perspectives"* (Malayalam), Papers presented in the International Conference on Democratic Decentralization (ICDD), Trivandrum, May 23-27, [eds.] Jagajeevan et.al.

———, (2003): Modernizing Government Programme [MGP], Theme V, Effective, Efficient and Accessible Local Self Government, Schematic Outlines of Detailed Implementation Plan, pp. 34-35.

Gupta, I.D (2002): "Corruption and Rent Seeking in Les Developed Countries: An Indian Perspective", *Indian Journal of Economics*, No. 328, July, Vol. LXXXIII, pp. 75-93.

Heidenheimer, Arnold J., Michael, Johnston and Victor, T.Le Vine (eds), (1989): *Political Corruption: A Handbook*, New Brunswick, NJ: Transaction Books.

Huther, Jeff and Shah, Anwar (1998): Applying a Simple Measure of Good Governance to the Debate on Fiscal Decentralization, World Bank Policy Research *Working Paper* 1894, Washington, D.C.

Institute of Social Sciences (2004&2005): '*Panchayati Raj Update*', various issues.

John, M.S and Chathukulam, Jos (2001): "Social Capital Formation and Institutional Growth Following Participatory Planning in Kerala: A Critical Appraisal", in *Emerging Institutions for Decentralized Rural Development*, [ed] S.P. Jain, Proceedings of NIRD Foundation Day Seminar, January 7-8., Vol. I, NIRD Hyderabad, pp. 478-489.

Kohli, Suresh (1975): *Corruption in India*, New Delhi, Chetan Publications.

Krishnakumar, R [2005]: "Power to the People Eroded", *The Hindu Daily*, January 2, p. 14.

Leys, Colin (1989): "What is the Problem about Corruption", in *Political Corruption: A Handbook,* [eds] Arnold J. Heidenheimer., Michael Johnston, and Victor T., Le Vine, New Brunswick, NJ: Transaction Books, pp. 51-66.

Mauro, Paolo (1995): "Corruption and Growth", *Quarterly Journal of Economics,* Vol. 110, August, pp. 681-712.

——— (1997a): "Why Worry About Corruption?", *Economic Issues*–6, International Monetary Fund, Washington D.C., p. 1.

——— (1997 b): "The Effects of Corruption on Growth, Investments and Government Expenditure: A Cross Country Analysis", in *Corruption in the Global Economy,* [ed.] Kimberly Ann Elliott, Institute for International Economics, Washington.

McLean, Keith (2001): *"Political Decentralization: International Experience",* Paper presented in the International Conference on Democratic Decentralization in Kerala: Status and Next Steps, Trivandrum, August 17-18.

Mello, Luis de and Barenstein, Matias (2002): "Fiscal Decentralization and Governance: A Cross Country Analysis", in *Governance, Corruption and Economic Performance* [eds] George T. Aded and Sanjeev Gupta, IMF, Washington D.C., pp. 333-365.

Mohanan, B (2000): "Controlling Corruption at the Grossroots: Lessons from a Decentralized and Participatory Development Administration" in *Corruption at the Grassroots: The Shades and Shadows* [eds.] Narayanaswamy, M.P. Boraian and M.A. Jayaraju, Concept Publishing Company, New Delhi, pp. 116-136.

Moody–Stuart G (1994): *Grand Corruption in Third World Development,* Transparency International, Berlin.

Nye, Joseph S (1967): "Corruption in Political Development: A Cost-Benefit Analysis", *American Political Science Review,* 61, June p. 417.

Oommen, M.A (2004): *Decentralization and Poverty Reduction,* UNDP Study as reported in Economic Review 2004. Government of Kerala, p. 466.

Pavarala, Vinod (1996): *Interpreting Corruption: Elite Perspectives in India,* Sage Publications, New Delhi.

Prudhomme, Remy (1994): On the Dangers of Decentralization, Policy Research *Working Paper*–1252, Transportation, Water and Urban Development Department, The World Bank, Washington D.C.

Rajasekharan, K (2004): "Citizens' Charter: A New Tool for Vitalizing Local Government Services in Kerala", *Panchayati Raj Update,* Vol. XI, No.10, 130, October, p. 8.

Sankaran, P.N and Vijayakumar, B (2001): "Proliferation and Interface of Institutions Under Decentralized Planning: The Kerala Experience", in *Emerging Institutions for Decentralized Rural Development,* [ed] S.P Jain, Proceedings of NIRD Foundation Day Seminar, January 7–8, Vol. 1, NIRD Hyderabad, pp. 446–464.

Socio Economic Unit Foundation, Kerala (2001): "Workshop on Construction Works under the People's Planning Campaign, Kerala: Social and Community Contracting Issues", *Background Papers,* 24 May, Trivandrum.

Sreekumar, G.S and Gopakumar, V (2005): 'Citizens' Charter: A Tool for Social Auditing, Paper presented in the International Conference on ' A Decade of Decentralization in Kerala: Issues, Options and Lessons' organized by the Institute of Social Sciences, 7-9 October, Thiruvananthapuram, Kerala.

Stockmayer, Albrecht (1999): "Decentralization: Global Fad or Recipe for Sustainable Local Development?", *Agriculture and Rural Development,* Vol. 6, No. 1, April, Chapter on Decentralization, p. 3.

Tanzi, Vito and Hamid, Davoodi (1997): Corruption, Public Investment and Growth, IMF *Working Paper* 97/139, Washington, IMF.

———(1998): Roads to Nowhere: How Corruption in Public Investment Hurts Growth, *Economic Issues* – 12, International Monetary Fund, Washington D.C, p. 10.

Tanzi, Vito (1998): "Corruption Around the World: Causes, Consequences, Scope and Cures", *IMF Staff Papers*, Vol. 45, December, pp. 559–594.

The Economist (1991): "*A Survey of India*", May 4, pp. 3-18.

United Nations Development Programme (1998): *Human Development Report*, Oxford University Press, New York, p. 11.

Vijayanand S.M (2001): "Issues Related to Administrative Decentralization and Administering Decentralization: Lessons from the Kerala Experience", paper presented at the Workshop on Decentralization, Institute for Social and Economic Change, Bangalore, 31 May and 1 June, p. 14.

Wolf, Thomas and Gurgen, Emine (2000): Improving Governance and Fighting Corruption in the Baltic and CIS Countries: The Role of the IMF, *Economic Issues*–21, International Monetary Fund, Washington D.C., p. 2, 7.

World Bank (1990): *Peru: Sector Reform and Investment Review*, Technical Department, Latin America and the Caribbean, The World Bank, Washington D.C.

———(2000): *Rural Decentralization in India*, Part III, The World Bank, Washington D.C. pp. 105–136.

8

LOCAL SELF GOVERNMENTS AND EDUCATIONAL DEVELOPMENT IN KERALA

INTRODUCTION

Social sector refers to those activities, which contribute to human capital formation and human development. Education, health and medical care, housing and sanitation, drinking water, etc are the main factors which contribute in the process of development of human capital or for that matter in human development (Panchamukhi, 2000). The issue of development of the social sector has acquired increasing importance in recent times. It is being increasingly recognized that economic development, or increase in per capita income of the population should be adequately matched by development of social sector if the process of development is to lead to increase in welfare of the common man (Pant and Pandey, 2001).

In the context of Indian planning, the development of social sector, particularly education, has always been an area of prime concern for planners, social activists and academicians and has therefore figured at the core of five year plans. The impact of planned development efforts, however, became more conspicuous on economic sector than on the social sector. In the case of education and literacy, various field level studies show that though there has been an improvement in the overall level of literacy, number of schools, students enrolled and teachers (Mehta, 1995), but so has been the level of drop-outs of students at the primary and middle levels (Majumdar and Vaidyanthan, 1994), increasing marginalization of socially disadvantaged groups, dilution in the quality of education (Pillai, 1992; Visaria, Gumber and Visaria 1993) and the high proportion of illiterates. Similarly, progress in the health sector (lowering of crude birth rate, death rate, infant mortality rate, enhancement in the life expectancy at birth etc.) has, by and large, remained confined to urban areas only.

Eventhough under the reforms package, the role of the state has been redefined, social sector development continues to be one of the main agenda of the government. The link between the expenditure on various components of the social sector and the growth of GDP has been examined by Abdus Salam (2001) for the reforms period 1991-92 to 1999-2000 in terms of a regression analysis taking GDP as dependent variable and components of social sectors as independent variables. Results indicate that the major portion of growth in GDP has been due to the growth in expenditure on social services as a whole. Empirical results broadly indicate that expenditure on social sectors in general and on education and social welfare in particular have contributed more in the growth of GDP.

The devolution of development functions, institutions, personnel and finance, following the 73rd and 74th Amendments to the Constitution of India, has opened up new vistas of social sector intervention for Local Self Governments (LSGs) across the country. In states like Kerala, a pioneer in decentralized participatory planning, 35-40 per cent of the plan funds of the state are allocated to local governments for preparation of locally relevant development plans for economic growth and social justice. The economy of Kerala has achieved rapid strides in the distribution of two major public goods viz education and health. Development experience of Kerala in the social sector has been comparable to international standards; however, the sustainability of social sector development has been open to doubt as the state has been experiencing stagnation in production sectors and a low level of per capita income, compared to other states. In the above context, the development paradigm of Kerala needs to be reexamined in the light of participatory intervention of local self-government in production sector as also the social sector. The present paper has been conceived against the above background of LSG intervention in education over the Ninth Five Year Plan period of Kerala.

DECENTRALIZATION AND EDUCATIONAL DEVELOPMENT: AN OVERVIEW

School decentralization arguments have been widely adopted in the rhetoric of education reform in many countries around the world (Peru 1972; Philippines 1974; Nigeria 1977; Chile 1980) particularly in the English-speaking States. Grant (1979) takes a incisive look at how the schools and the community can work as partners in making important school policy decisions in areas such as budget, curriculum planning, selection of school personnel, and plans for racial integration. In a thought-provoking review of the school "Production function" literature, Hanushek (1986) finds no relationship between increased resources and improved results in schools. He concludes that schools do not use resources efficiently.

Winkler (1989) has developed a Centralization-Decentralization Typology for public education (Table 1 and Chart 1) by disaggregating the provision of educational services in to its various components or functions viz school organization, curriculum and teaching methods, examination and supervision, teacher recruitment and compensation, finance of recurrent expenditures, and school construction and finance. It is the combination of decision- making authority with respect to the various functions, which determines whether the system can be characterized as centralized or dicentralised. But the answers to three questions best discriminate between centralized and decentralized systems. They are:

Table 1
Centralization-Decentralization Typology for Public Education

	School Organization	*Curriculum and Teaching Methods*	*Examinations and Supervision*
Centralized Model	Minimum schooling requirements and school organization(preschool, primary, secondary, vocational, higher education) set by the central government.	Curriculum teaching materials, pre-service and in-service instruction provided by the central ministry of education	Examinations set and evaluated, as well as teaching performance evaluated by central ministry of education, responsibility for direct supervision often lies with regional administrative offices.
Mixed Model	The central government determines organization of the schooling system, but the local community helps determine how many years of education are provided, often through at least temporary self-finance of years beyond those funded or authorized by the central government.	Curriculum, teaching materials, and in-service instruction established and provided by the central government or through its regional delegations.	Examinations are set centrally but usually administered and evaluated regionally; the instruction, often through regional or district offices.
Decentralized Model	Organization of schooling is almost always set by the central ministry of education; the local community decides how many years and level of education will be provided.	The basic contents of the curriculum are set centrally, but textbooks may be selected and purchased locally, and in service instruction may be provided locally or regionally depending on the size of the locale.	No national examination system exists; all examinations are set and evaluated locally; the central or regional governments usually provide limited supervision of teachers and schools.
	Teacher Recruitment and Compensation	**Finance of Recurrent Expenditure**	**School Construction and Finance**
Centralized	Central Government sets accreditation standards, provides teacher education, sets teacher pay scales, and directly pays the teachers; in some cases (Franco-phone Africa) teacher recruitment, pay and promotion may be under control of the civil service ministry rather than the ministry of education.	All recurrent expenditures fully funded by central government excepting minor user fees, non- teacher resources distributed to schools.	Central government sets construction standards, which may be uniform for the entire country, and covers all construction costs, although the local community may be required to provide labour and or some construction materials.

	School Organization	*Curriculum and Teaching Methods*	*Examinations and Supervision*
Mixed Model	Teachers may be selected by the local school authority, but the central or regional government typically prescribes pay scales; accreditation standards are also set centrally.	The Central or regional government provide most funding of local schools in the form of block grants or project grants, but some portion of educational expenditures are funded by local revenue sources, and the local community has some influence on total expenditure levels.	Construction standards are set by the central or regional government and matching funds are often provided for school construction; in some cases the matching funds take the form of a promise by the central government to cover some portions of recurrent expenditure, often teacher salaries.
Decentralized Model	Teachers are selected and pay scales are set by local government; accreditation standards are typically set by the central government but they may not be enforced.	Local government funds elementary and sometimes secondary education from local revenue sources; user fees or "voluntary" contributions to the parents-teachers associations may be required; block grants or project grants may be provided by the central government.	Land and materials for school construction are provided by the local community; labour may be voluntary; local construction standards used.

Source: Winkler Donald R. (1989), Decentralization in Education; An Economic Perspective, Policy, Planning and Research Working Paper, Education and Employment, Population and Human Resources Department, The World bank, WPS 143, p. 8-9.

Chart 1
Educational Decentralization: A Synoptic View

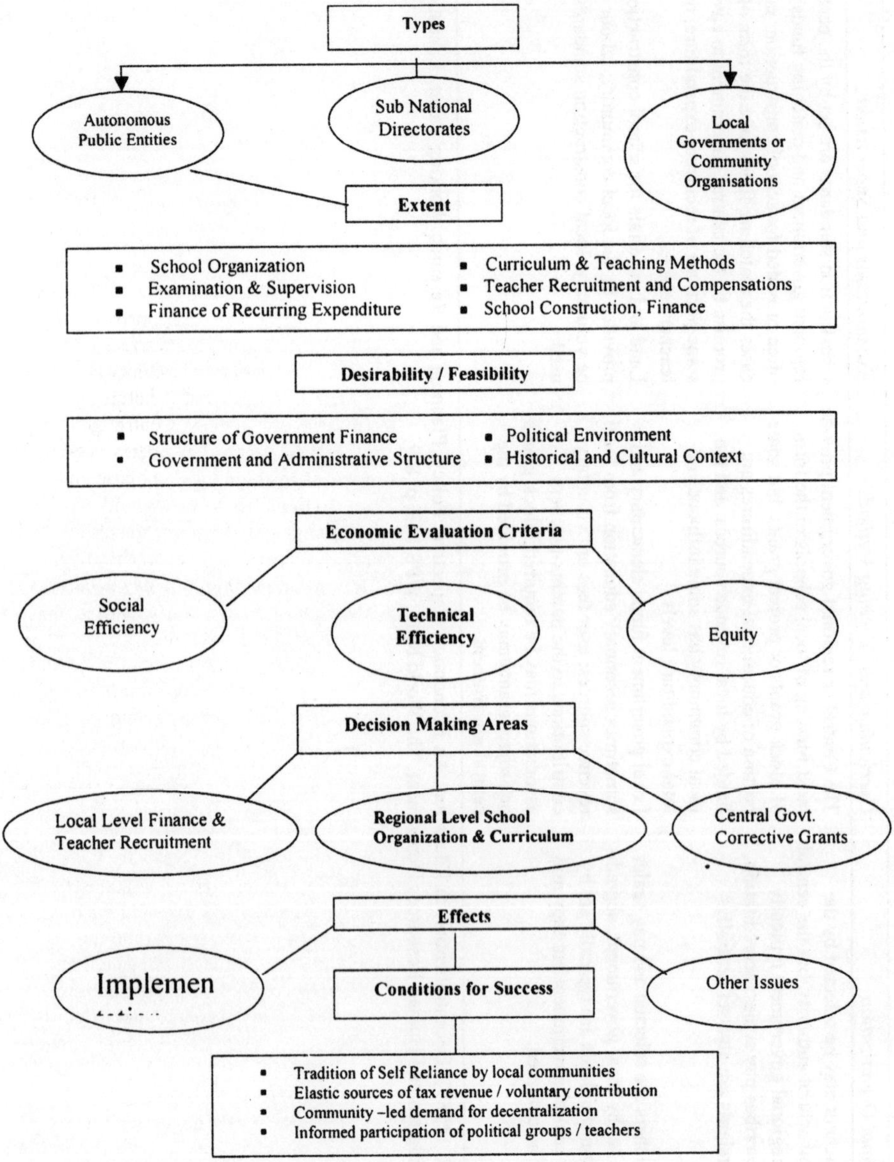

Source: Developed from Winker D.R. 1989, Decentralisation in Education: An Economic Perspective, The Word Bank, WPS, 143.

- Who selects the chief administrative officer of local schools, and what control do they have over resource allocation?
- Which level of government is responsible for recruiting and promoting teachers?

- What proportion of total expenditures is financed through local revenue sources, both tax revenues and voluntary contributions?

Three economic criteria are used by Winkler to evaluate decentralization—*social efficiency, technical efficiency and equity*–(Chart 2). The rationale for educational decentralization has been grouped into three broad categories.

Chart 2
Economic Criteria for the Evaluation of Decentralisation in Education

Criteria	*Definition*	*Implications for Centralization/Decentralization in Education*
Social Efficiency		
Public Choice	The match between consumer-citizen preferences and public service provision	Local citizens should have an important voice in choosing the combination of taxes/fees and educational services in the community and should also, have some influence on the nature of educational services provided. Central govt. regulations, restrictions and mandates may limit public choice, as may the iocal tax base and impediments to the citizen "voting with his feet".
Externalities	The discrepancy between who pays for services and who receives the benefits	Inter jurisdictional spillovers may result in insufficient educational spending in a decentralized system. The central government may need to institute a system of matching grants to assure a socially desired level of spending.
External efficiency	The balance between labour skills provided by the educational system and labour market demand	In a country with distinct urban/rural and regional labour markets, there should be variation within the country in terms of school organization, curriculum and the quantity of schooling.
Technical Efficiency		
Internal Efficiency	Maximization of educational output for a given educational cost	Internal efficiency is higher when prices of educational inputs are allowed to vary with local(urban/rural, regional) market conditions, and the educational input mix is adjusted for local prices. Economies of scale may argue for centralization of some educational functions. Administrative costs, including both administrative capacity and the time required to make decisions, also affect unit costs and , thus internal efficiency.
Technological Change	An increase in the educational output possible from a given level of inputs	Technological change and innovation in educational organization and service delivery is likely to be greater in a system which both permits and encourages educational diversity and competition.
Equity		
Horizontal Equity	Individuals in like circumstances are treated equally	When some portion of school revenues are generated from local tax sources, variations in wealth of the local tax base may lead to horizontal inequity. The central Govt. can reduce the size of the problem through grants in aid to local school with the size of the per pupil grant varying inversely with the local tax base per capita.

Criteria	*Definition*	*Implications for Centralization/Decentralization in Education*
Vertical Equity	Relative equality of educational opportunity	Equality of educational opportunity may be measured by access to and quality of education. In centralized systems, inequality is the result of central government decisions regarding resource allocation; in decentralized systems inequality may result from variations in local tax base and citizen willingness to pay for education. In either case the central govt. may use grants in aid to reduce spending disparities and improve equality of opportunity.

Source: Winkler Donald. R (1989) Decentralisation in Education; An Economic Prospective, Policy, Planning and Research Working Paper, Education and Employment, Population and Human Resource Department, The World Bank, WPS 143, P17.

(i) educational finance (ii) efficiency and effectiveness, and (iii) redistribution of political power. In view of fiscal constraints of state governments, shifting part of the burden for support of primary and secondary education to local government, to community and voluntary organizations and to parents has become an increasingly attractive alternative. Administration and accountability can be improved in education, by making schools more responsive to parents and local communities. Though rarely stated as an objective, redistribution of political power involves democratization/inclusion of marginal groups in educational decision-making.

Drawing on theoretical models, case studies and comparative analyses of other sectors and countries, Hanaway and Carnoy (1993), argue that governance reforms are likely to have little impact on what actually happens in schools. And they tell why decentralized structural arrangement alone is unlikely to establish conditions necessary for general improvement in educational practice.

Among economists, advocates of efficiency gains associated with decentralization in the education sector typically rely on two hypothesized mechanisms: [1] competition for students can improve schools so that school choice for users is an important institutional feature, and [2] decentralization can make the education system more nimble by exploiting voice and participation, either at the level of local government or directly at the school level, as autonomy is expanded at lower levels in the system and ultimately to the school level.

There is not much hard and quantitative evidence on whether decentralization per se can improve the efficiency of schools through voice and participation. Jimenez and Sawada [1999] find that participation by parents and the local community improves performance in EI Salvador's EDUCO schools, despite the much poorer conditions in which they work.

In the context of Argentina, Eskeland and Filmer (2002) has analyzed the relationship between autonomy, participation and output by estimating a traditional education production function of the form.

$$t = f[X_1, X_2, A, P], \text{ where}$$

- 't' is test scores in the national evaluation system;
- X_1 is a vector representing the socio–economic background of students, to represent private inputs into the production process;
- X_2 is a vector representing school inputs (for example the education of teachers).
- A, P is a vector of variables representing autonomy and parental participation at the school level. The conclusion is that there is interaction of autonomy and participation matters for learning (in mathematics).

According to their study, a positive effect on learning attributed to participation and autonomy at the school level is relevant to the question of decentralization in two ways: [i] if decentralization leads to more autonomy and participation at the school level, then the results appear to be relevant without further assumptions, and [ii] decentralization typically would mean more autonomy and electoral participation at lower levels of government, and the results are relevant if one assumes they reflect more general effects of bringing decisions closer to users or the community, geographically and jurisdictionally. Nevertheless, both interpretation require restrictive assumptions: it is not necessarily the case that general decentralization in the sector [say, transferring schools from provinces to local government] leads to more autonomy and participation at the school level, nor is it necessarily the case that bringing decision making closer to the school [say to, local government] has effects similar to decision making at the school.

To sum up, studies revealing that schools are far from efficient in translating resources into learning [for example Hanushek 1986, 1994, 1995] have led to an increasing interest in the education literature on issues such as governance, reform and decentralization.

Claims that decentralization will enhance the effectiveness of school systems have formed the central plank in the works of many prominent school reform advocates [Chubb and Moe 1990: Sexton 1987]. However, comparative studies have not adequately established that it can achieve higher trends of 'value added' than the supposedly more centralized schools in the public system [Carnoy, 1993]. According to Andy Green [1997], available data suggest that one group of countries [France, Germany, Japan, Sweden, Singapore], whose education systems have been traditionally characterized a 'centralized' [Archer 1979; Boucher 1982; Gopinathan 1994; Green 1990; Schoppa 1991] appear to obtain better aggregate attainments than another pair of countries [UK and USA], which are traditionally characterized as 'decentralized' [Archer 1979, Green 1990, Kaestle 1983]. Witte [1990], writing on a US context, concludes that arguments for and against decentralization are difficult to sort out and that there is as yet little evidence about the impact of decentralization on results. However, the above results emerging from small sample of countries, does not have universal application on the relative merits of centralization / decentralization in educational system.

There are, of course, studies indicating a positive relation between school resources and schooling outputs [Card and Krueger 1992]. Edited volumes such as Clune and Witte [1990], Hannaway and Carnoy [1993] and Chapman and Others [1996] are evidence of the interest in the question whether institutional reforms such as decentralization lie at the heart of school's problems.

Decentralization experience reviewed above shows that there are serious constraints on the ability of local governments or communities to raise revenues to support local schools, improve efficiency in the delivery of educational services and minimize inequity in the system.

EDUCATIONAL DECENTRALIZATION

(a) India

Association of local bodies with primary education is not a new phenomenon in India. According to Premi [1998], the beginning of the era of local self-government including local administration of primary education was registered with the historical resolution [1882] of Lord Ripon on local self-government. The Montagu Chelmsford Reforms and subsequently the Government of India Act [1919] accelerated the process of transfer of powers to local bodies. The Primary Education Acts passed by the Provincial governments during the 1920's provided much larger authority to local bodies over primary education. The period from 1929–47 was dominated by attempts to curtail the powers of local bodies in education. Article 40 of the Constitution of India gave positive signal for people's participation in education through village panchayats. The Study Team on Community Development headed by Balwant Rai Mehta wanted education at the lower level to be handled by local bodies. However, the Education Commission [1964–66] was critical of the approach and recommended withdrawal of village panchayats from educational interventions. The Report of the Ashok Mehta Committee [1977], on Panchayati Raj Institutions was not able to revive the association of village panchayat in basic education. Most of the states, therefore, witnessed a marked decline in the role and power of Panchayati Raj Institutions in respect of school education in the eighties. The National Policy on Education [1986] emphasized the importance of decentralization in planning and management of education at all levels by ensuring community participation. It may, however, be observed that the real impetus to educational decentralization was given by the PR Acts passed by state governments following the 73rd and 74th Amendments to the Constitution of India in 1992. The CABE committee for formulation of guidelines on the decentralized management of education in the context of 73rd and 74th Constitution Amendment Acts, envisaged an important role for LSGs in exercising supervision over early childhood care and education, non–formal, primary and adult education in their jurisdiction. The 73rd Amendment to the Constitution has specifically transferred education to the area of operation of the panchayat in the Eleventh Schedule [Article 243 G] in terms of the following items:

- Education, including primary and secondary
- Technical training and Vocational education
- Adult and non–formal education

- Libraries, and
- Cultural activities

Consequent to this amendment, the states have legislated the necessary amendments to their respective PR Acts, covering the areas mentioned in the Eleventh Schedule. CABE recommended that a Panchayat Standing Committee on Education [or the Village Education Committee as a sub committee of panchayat] may be constituted at this level. The role of the committee may include generation and sustenance of awareness among village community, ensuring participation of all segments of population, developing teacher and community partnership to oversee and to supervise effective and regular functioning of the schools and centers.

(b) Kerala

Kerala's social development is unique that it has successfully tackled the first generation problems–illiteracy, high infant mortality rate, high birth rate and related indicators of under development–while other Indian states are still grappling with the first generation problems. This has been primarily due to the rapid strides made in extending the two major public goods namely, education and health. The plan expenditure of Kerala on social services constituted about 24 per cent of the total outlay from the first to the Ninth Five Year Plan whereas that all India figure ranges between 17–20 per cent. The non–plan expenditure on social services constitutes about 36 per cent of the total government expenditure of the state [GOK, 2001].

Infrastructure of school education in Kerala during 2000 - 2001 consisted of 12330 schools split up into LP [6758], UP [2957] and HS [2615]. For every 34000 persons there is an LPS/section; for every 236 children in LP age group, there is one LP school/section. There is one UP school/section for every 329 UP school going age group population. The spatial distribution of schools shows that there were 651 schools in the Corporation areas comprising of 228 LP schools, 144 UP schools and 219 High schools. In Municipality areas, there were 932 schools that constituted of 457 LP schools, 207 UP schools, 268 High schools. The remaining 10727 schools are in Panchayat areas. Of these, 4724 are under District Panchayats and 6003 under Grama Panchayats.

The vast educational infrastructure facilitates easy access to LP, UP and HS education in Kerala. It is estimated that 94.39 per cent of the rural population is served by primary schools within a distance of 1 km and 97.96 per cent within a distance of 2 km. In regard to UP education, 96.20 per cent of the population is served with a UPS within a distance of 3 kms. The rural population served by secondary schools at a distance up to 2 kms is 24.71 per cent and 97.82 per cent has the facility for secondary education between 6 and 8 kms. Schools in Kerala have better facilities compared to those in other parts of the country as revealed from the Educational Survey Report of NCERT. In 2000 about 83 per cent of the schools were housed in proper building with basic amenities, 84 per cent of the Government schools have drinking water and 85 per cent have latrines/urinals.

However, the NCERT study shows that the position of Kerala is only 17 among the states in terms of the quality of school education.

The Kerala Panchayati Raj Act/Kerala Municipality Act 1994 provide that Primary schools will be managed by Grama Panchayat and UP and High schools will be managed by District Panchayats, Municipalities and Corporations. Higher secondary schools in the Government sector are also transferred to the LSGs at appropriate levels. Over and above the state sector projects planned and implemented towards quality improvement and basic facilities in schools, the LSGs formulated 10520 projects in the first two years of the 9th Five Year Plan and allocated Rs. 64.17 crores. The projects of the LSGs mainly aimed at improving amenities in schools, construction of class rooms, compound walls and improvement in quality of education through special efforts like social mobilization, parental education, training of teachers and so on [GOK, 2000]. In the effective implementation of the centrally sponsored "Sarva Siksha Abhiyan" [SSA] also peoples' participation through LSGs is envisaged.

LSG INTERVENTION IN THE SCHOOL SYSTEM: AN EVALUATION

As already stated, the focus of the paper is an evaluation of LSG intervention in school education over the 9th Five Year Plan of Kerala. In view of the above objective, we have made:

(a) An evaluation of the educational intervention/projects of Thiruvananthapuram Municipal Corporation in Kerala during the 9th Five Year Plan period [1997–02] in terms of secondary sources of information, and

(b) An analysis of the response/opinion of a cross–section of people formally or informally involved in the process of educational intervention in a selected Grama Panchayat in Thiruvananthapuram District, Kerala, with reference to Annual Plan 2001–02 of 9th Five Year Plan period, using the survey method.

With a view to generate further issues for discussion, research and action, an attempt is also made to scan the intervention of Thiruvananthapuram District Panchayat in the education sector during the first year [2002–03] of the 10th Five Year Plan period.

The review of the school sector in the educational system of Kerala has indicated that it ranks high in terms of infrastructure and enrollment; however, the quality of learning/teaching as also the rate of drop out leave much to be desired. In this context, a glance at the existing arrangements for quality improvement in the state seems appropriate. At the state level, Kerala has the State Council for Educational Research and Training [SCERT], an autonomous body with the declared objective of preparation of curriculum, imparting training to teachers and conducting research activities on issues pertaining to school education up to 12th standard. At the District Level we have District Institute for Education and Training [DIET] in all districts, meant for imparting training to primary school teachers. Besides this, in order to give further thrust to quality improvement in primary schools, in all districts of Kerala District Programme Officers are appointed and under them Block Resource Centres are established. These centres are expected to give periodical training to primary school teachers for quality improvement. Hence it would be meaningful to evaluate the educational projects of the selected local body [Thiruvananthapuram Municipal Corporation] with reference to the above structure and the factors relevant to the study area/context.

From the analysis of the activity–wise intervention [rather than project–wise] of the Corporation in the education sector over the 9th Five Year Plan period as summarized in Table 2, we observe that 95 per cent of the total funds were utilized for construction of buildings. A review of the physical achievements in the school system during the 9th Plan period LSGs indicate the populist tendency to concentrate on infrastructure improvement. Over the 9th Plan period, LSGs have completed the construction of 180680 m. sq. of new schools built area under the General Sector [173923 M.sq], Special Component Plan [2208 M.sq], and Tribal Sub Plan [4549 M.sq] components [GOK, Economic Review, 2002]. An analysis of the educational investment pattern of LSG Annual Plan 2002–03 also endorse the above tendency. Available data pertaining to few districts [Kollam, Palakkad and Kasargod] shows that out of 1688 projects undertaken by the LSGs during 2002–03, 723 [43%] were envisaged for educational infrastructure development. In terms of the percentage share of plan grant/gross outlay also, the preference for infrastructure development is quite visible in the above districts. Projects directly relevant for improvement of the quality of learning/teaching covering teachers and weak students in particular, have received scant attention in the priorities of the Corporation's resource allocation strategy.

Hence it may be included that decentralization of governance and planning have by and large bypassed the real/basic issues in the school system namely quality improvement and equity; in fact the process of intervention has laid undue emphasis on more populist physical achievements in the sector, which seems to render more elastic political mileage to the elected functionaries of local governments. Projects addressing stabilization of the system in terms of reducing dropout at the high school level were also not covered by the planned participatory interventions of the Corporation. The basic scenario calls for more active and informed participation of the people, especially the educated, in the deliberations of Grama Sabha/working group on education and setting of priorities in local planning. Only by strengthening people's participation in the crucial stages of project identification, appraisal, implementation and monitoring, the will of the electorate will prevail. To sum up, the evaluation of the educational projects of the Corporation indicate the continuation of the traditional outlook to local planning rather than bottom–up participatory planning which is expected to address the basic issues in the local economy in terms of projects that are fine-tuned by the active/voluntary participation of experienced and skilled personnel of local bureaucracy/Grama Sabha.

Table 2
Educational Projects under the 9th Five Year Plan (1997-02): Thiruvananthapuram Municipal Corporation, Kerala

	Nature of Projects	*Amount utilised (Rs.)*
1.	Improvement of Laboratories in 27 U.P. Schools	7,77.320 (1.21)
2.	Construction of multi-storied building for 6 Higher Secondary and one UP School	3,50,00,000 (54.35)
3.	Improvement of Laboratory in one GHSS	15,00,000 (2.34)
4.	Supplementary coaching in 10 High Schools for improving levels of learning and pass rate of weak students	7,81231 (1.21)
5.	Establishment of Anganwadis and training of Anganwadi Preprimary teachers	80,000 (0.12)
6.	Improvement of Latrines in UP schools	54,281 (0.08)
7.	Construction of school buildings including an ITI	261,55,196 (40.61)
8.	Grant for improvement of co-operative societies in 5 schools @ Rs. 10,000	50,000 (0.08)
	Total	643,98,028 (100.00)

Figures in parenthesis indicate percentage.

Source: (1) Trivandrum Municipal Corporation (n.d) Evaluation of Ninth Five Year Plan and Approach to the Tenth Five Year Plan: Draft Document.

(2) Trivandrum Municipal Corporation (2002), Development Seminar: Tenth Five Year Plan (2002-07), Draft Project Document.

A MICRO ANALYSIS

Given the backdrop of national as well as international experience in decentralized decision-making, let us examine the intervention of LSGs in Kerala in the educational sector. Following the devolution of power in selected sectors, the High Schools were brought under the jurisdiction of the District Panchayat and UP and Primary Schools were brought under the control of the concerned grama panchayat with the declared objective of ensuring adequate infrastructure facility and maintaining proper quality standards. Keeping this broad objective in mind, this section makes an attempt to examine the nature and impact of intervention by LSG in school education focusing on Maranalloor Grama Panchayat, in terms of secondary information and an opinion survey.

STUDY AREA AND SAMPLE

Maranalloor Panchayat, located in Nemon Block of Trivandrum district in Kerala is spread over an area of 2484.35 hectares. With paddy and coconut cultivation as major economic activity, the panchayat depicts almost all features of a typical backward rural area. Though there are 10 schools in the government (5) and government aided (5) sector and one unaided school, both infrastructure facility and quality of education are not up to the mark. The total enrollment in all schools is 5717 with 2955 boys and 2762 girls. The total number of teachers is 183 with a teacher student ratio of 1:31. School education was given due weightage in the development agenda of Maranallor gramapanchayat. In order to elicit the opinion of concerned citizens on the planned educational intervention of the grama panchayat, a survey was conducted among the stakeholders of school education, [leaving the pupils], who comprise of the representatives of gramapanchayat, school teachers and the parents. The respondents were selected at random and the sample size constitutes 55. In the sample there are 30 [55%] parents, 15 teachers [27%] and 10 [18%] representatives of gramapanchayat. A questionnaire was designed to elicit their opinion on selected issues relevant to educational intervention of the gramapanchayat.

DISCUSSION

Without any exception, respondents were unanimous in their view that, following the introduction of Peoples Plan Programme in 1997, the panchayat was generous in promoting infrastructure facilities to schools by offering new buildings, repairing the old ones, providing furniture, toilet facilities and constructing compound walls to government schools.

However, the Ninth Plan evaluation report of the panchayat reveals that out of the 8 infrastructure development projects approved, under Annual Plan 2001–02, it has succeeded in completing only one and the remaining were either carried over to the next plan [5] or dropped midway [2]. While in 2 schools [20%] the decision to improve infrastructure facilities were taken by the school authorities, in 3 schools [30%] the decision was taken by the Parent Teacher Association and in the case of remaining schools [50%], the panchayat decided to go ahead with the construction of additional blocks, compound walls, toilet, repairing existing ones, making furniture etc. The activities of the local body are confined to the provision of physical facilities to the schools and not improvement in the quality of teachers/students.

Implications of [affirmative] opinion expressed by the sample respondents on selected areas of LSG intervention in school education are summarized in Table 3. The 73rd and 74th Amendment of the Constitution offer ample scope for the panchayat as well as the local community to take part in the development programmes at the grass root level. They can influence the decision-making process as well as work in the monitoring agencies of the ongoing economic activities. The community link to schools and devolution of power to panchayats are expected to improve the quality of education by giving due attention to neglected areas in the school education spectrum. Accordingly, we have identified three groups of issues for the opinion survey, namely infrastructure development quality improvement and other activities. The issues pertain to drop outs, evaluation system, improved facilities for laboratory, library and information technology, problems of economically backward students, satisfactory completion of effective teaching days, student discipline, teacher accountability, sports and cultural activities, work experience, cleanliness health programme, in–service training, parental awareness, social activity, academic discussion, utilization of panchayat library and local resource mobilization for assessing the participation and involvement of the stake holders in school education etc. Opinion of the stake holders reveal the following.

Table 3
Opinion of Parents, Teachers and Elected Representatives on Intervention of LSGs in School Education

Role of LSGs in Local Schools	*Positive Opinion [%]*		
	Parents	*Teachers*	*Representatives*
[a] Infrastructure Development			
Drinking water, Library, Laboratory, Computer facility, Furniture, Compound wall, Repair of buildings, Building construction, Urinals, Electrification, Stationery etc.	53	67	90
[b] Quality Improvement			
Teaching days, Syllabus completion, Student discipline, Teacher accountability/Disciplinary action, Teacher training, Prevention of drop-outs, Evaluation, Remedial coaching, Monitoring, Study tour, Exhibition, Competitions etc.	73	47	30
[c] Other Activities			
Extra/Co–curricular activities, Social service, Health awareness, Physical training, Strengthening PTA etc.	80	93	70

Source: Primary Survey.

Next to teachers, parents are the energizers of quality improvement. The opinion survey reveals that parents (73%) are the most forceful advocates of LSG intervention for quality improvement in schools. A silver lining in this regard is the good attendance reported of parents in meetings, where there is ample scope for presenting new ideas for quality improvement. LSG intervention for infrastructure development is evaluated only next to quality improvement by parents in the sample. The role of the panchayat in the promotion of extra/co-curricular activities in schools is held high by the sample parents.

Of the stakeholders, teaching community is the most important constituent that can make positive contributions to quality improvement in schools as they have direct contact with the students from 10 am to 4 pm on all working days. Without any exception, all teachers are aware of the changes taking place in teaching methodology and all of them have undergone periodical training also. Considering the parameters recognized for intervention in quality improvement, a lot of things can be done by the teachers without incurring additional time and finance. However, according to the sample teachers (93%) the most effective area of operation in the school system by the LSG is extra/co-curricular activities, social service etc; next to this they evaluate (67%) infrastructure development as the possible domain of LSGs. Therefore the need of the hour is to make the teachers increasingly aware of quality improvement and enable them to participate in it in collaboration with the LSGs.

The power devolution consequent on panchayat raj and Peoples Plan gave local bodies scope for meaningful intervention in school education. But the entire scenario remain unchanged as an absolute majority (90%) of the representatives want to make only visible/physical changes during their tenure. Generally the parents, teachers, and elected representatives believe that henceforth it is the duty of the panchayat to provide basic facilities to schools based on the requirements placed by them. Practically no fruitful discussions are taking place in the Panchayat Education Committee and Academic Committee in preparing projects for improving the quality of school education. Elected representatives also want to show off the society that significant development of physical infrastructure has taken place during their tenure which will stand as testimony for their contribution to the concerned ward/panchayat. While the representatives of schools and parents seldom attend the concerned panchayat meeting, the elected representatives in turn refrain from the day to day business on the presumption that they have no 'right' (power) to take disciplinary action against dereliction of duty. The teachers believe and the representatives presume that, in the given context, elected representatives have little role to play in quality improvement programmes of schools which will have a long bearing on the younger generation.

The need of the hour is to create awareness among the stakeholders and enable them to identify ways and means for true participation in quality improvement rather than in cosmetic solutions. Quite often construction activities are undertaken without any assessment of future demand. It is high time for the panchayats to involve in quality improvement by making use of their physical and financial resources to accomplish the desired goals of education. A reorientation of the field of activities in the participatory approach, together with devolution of more power to the panchayat can cherish this goal.

APPROACH TO EDUCATIONAL DECENTRALIZATION: THE TENTH PLAN PERSPECTIVE

During the 10th Five Year Plan (2002-07) LSGs in Kerala propose to deepen their intervention in the school system in terms of 'Comprehensive Education Programme' (CEP) encompassing State and Centrally sponsored programmes including SSA (Chart 3). Acccording to the 10th Five Year Plan Approach Document (12) Education Sector, LSGs-2002 the objective of educational intervention in the Plan period is to enhance peoples' participation, secure equity and improvement in the quality of education output, improve capacity building in LSGs and establish sustainable linkages in the CEP with State and Central level projects and schemes implemented in the past without much coordination or as part of a broad strategy. Schemes of agencies/department like Health, Sanitation, Literacy, Scheduled Caste/Tribe, Cooperatives, Banks, Cultural Institutions, NGOs etc, need proper coordination with CEP. The proposed strategy takes an integrated approach to educational development at various levels, in which the LSGs down the line have to play a crucial role (Chart 4), keeping the 'pupil;' as the focal point of educational planning and intervention (Chart 5).

Chart 3
Comprehensive Education Programme

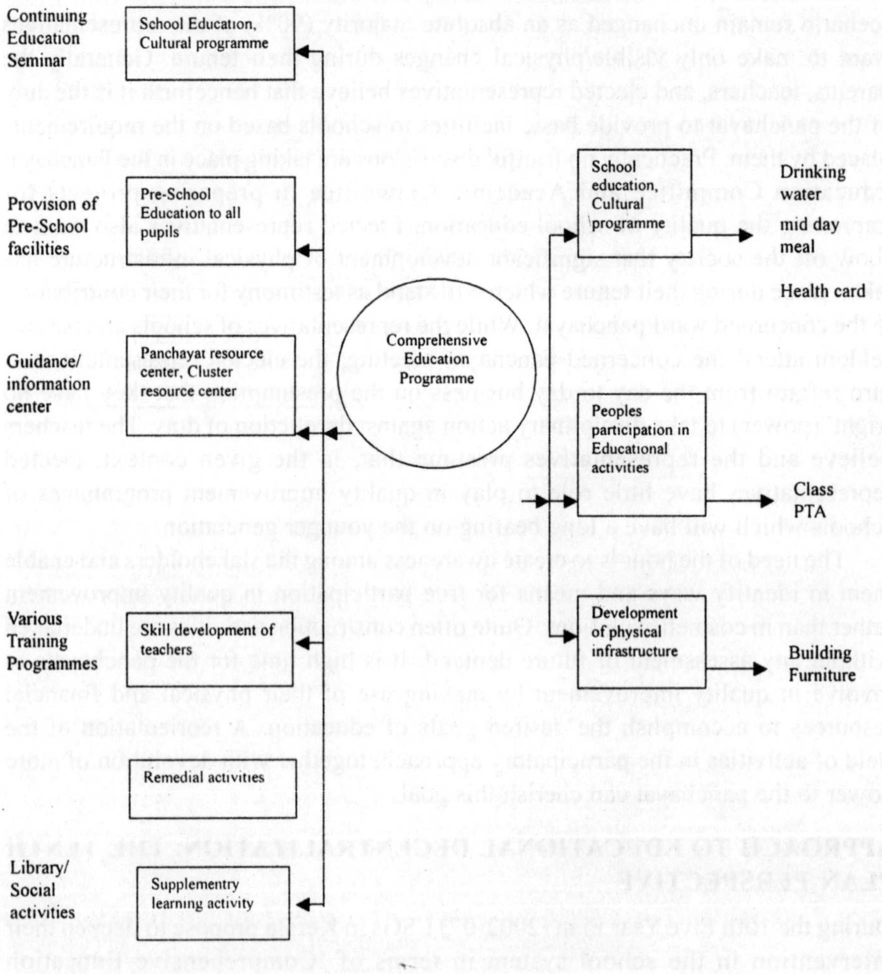

Source: Government of Kerala, State Planning Board (2002) Planning by Local Self Governments: Tenth Five Year Plan 2002- 2007, Health Sector, p. 23.

Despite the integrated and comprehensive approach envisaged in the educational intervention of LSGs during the 10th Five Year Plan period, the Annual Plan (2002-03) proposals of the Thiruvananthapuram District Panchayat (summarized in Table 4), do not make much radical departure from the infrastructure-oriented resource allocation pattern followed in the past towards educational projects. Allocation of resources for upgradation of school premises and buildings, however, has come down (30.34%); 21.04% of the resources have been set apart for computer based projects, which of course may work towards quality improvement in teaching and learning. Though, for want of comparable

Chart 4
LSGs and Educational Planning: The Linkages

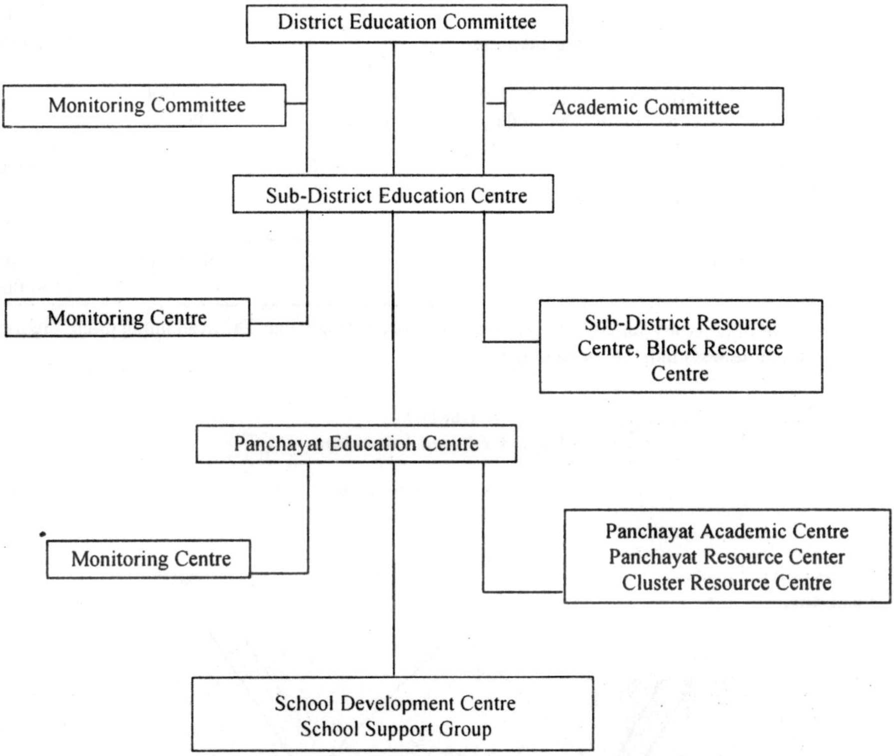

Source: Government of Kerala, State Planning Board (2002) Planning by Local Self Governments: Tenth Five Year Plan 2002–07, Health Sector, p. 20.

Table 4
Educational Projects (2002– 03) in the General Sector, Tenth Five Year Plan, District Panchayat, Thiruvananthapuram, Kerala (Amount in Rs.)

	Nature of Projects	*State Fund*	*Plan Fund*
1.	Quality improvement of weak students	-	4,67,000 (6.66)
2.	Beautification of School Premises	-	14,95,000 (21.33)
3.	'Gandhidarsan' is schools	-	4,55,000 (6.49)
4.	Establishment of Library / Laboratory in 10 High Schools	-	5,05,000 (7.22)
5.	Internet in schools and Net/computer connection in Educational Institutions with District Panchayat	-	7,20,000 (10.27)
6.	Computer installation in Schools	-	7,55,000 (10.77)
7.	Infrastructure improvement in TGMT centers and Commercial Institutes	-	1,00,000 (1.43)

Contd....

	Nature of Projects	*State Fund*	*Plan Fund*
8.	Physical Training	-	2,80,000 (4.00)
9.	Comprehensive Literacy in coastal areas	-	15,00,000 (21.40)
10.	Assistance for furniture making in 25 High Schools	7,55,000	
11.	Emergency relief to schools affected by natural calamities	2,00,000	-
12.	Encouragement of Arts	-	2,00,000 (2.85)
13.	Thatching of school buildings	-	5,31,000 (7.58)
	Total	9,55,000	70,08,000 (100.00)

Source: Tenth Five-Year Plan 2002-07, Proposals, District Panchayat, Thiruvananthapuram. Figures in parenthesis indicate percentage.

Chart 5
Pupil-Centered Education

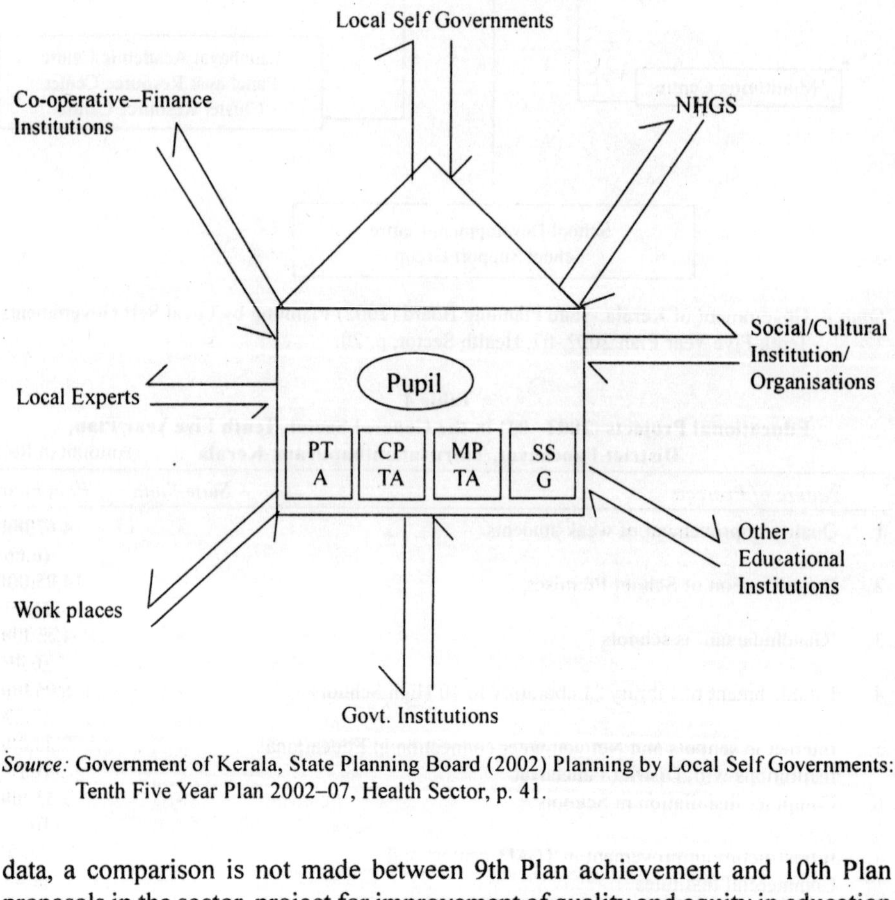

Source: Government of Kerala, State Planning Board (2002) Planning by Local Self Governments: Tenth Five Year Plan 2002–07, Health Sector, p. 41.

data, a comparison is not made between 9th Plan achievement and 10th Plan proposals in the sector, project for improvement of quality and equity in education need greater priority in the participatory process of LSG intervention in school education. The gradual shift towards diversified and quality linked projects noted above also do not by itself reflect the emerging trend at the macro level.

CONCLUSIONS

Kerala is acclaimed as a pioneer in achieving significant progress in social sector development among all other states in India. The development indicators of the state, comparable with some of the developed countries of the world, are partly due to the achievement in social sector. But a perusal of what is achieved in school education would reveal that the accomplishment is more in quantity than in quality. The low ranking of Kerala at the 17th position by the evaluation study of NCERT substantiates this statement. The pertinent question in this context is whether the invention of LSGs in social sector can make positive contribution to quality improvement in school in Kerala.

The secondary information and field survey reveal that in spite of Peoples Plan, the LSGs by and large follow the conventional path for executing their development programmes. An overview of the various development projects accepted by the LSGs show the overemphasis given to physical infrastructure rather that peeping into the qualitative aspects. The social sector is not an exception. A shift in the line of approach necessitates creation of better awareness among all stakeholders on participatory approach. Though Kerala has completed few years of participatory planning, the programme is in its infancy as the state is passing through a transition period from conventional planning to participatory planning. A synchronization of stakeholders with participatory approach necessitates good groundwork at grassroot level in which building awareness among all concerned is the focal point of action.

In the light of the mushroom growth of unaided schools, division-fall in government and aided schools, closure of uneconomic schools etc, the state has to take earnest efforts to maintain quality rather that giving undue emphasis on prompting the creation of physical infrastructure. As already mentioned, the state has a well-knit mechanism for imparting periodical training to school teachers for quality improvement. Besides, the agencies involved are responsible for periodical changes in syllabus and publication of study material also. But there is no effective mechanism for evaluating the progress achieved in quality improvement following the training imparted to teachers. As the LSGs have easy access to schools and the elected representatives can establish better rapport with school authorities, the LSGs can think about devices whereby proper evaluation of the progress in quality improvement in schools coming under their jurisdiction can be effectively undertaken. The LSGs can place their evaluation repot to training institutes/ agencies so that the latter can frame future courses of action for better results. A switchover to quality aspects through participatory techniques envisages good homework, besides strong will of all stakeholders. Steps may be taken to induct quality aspects in the broad framework/guidelines given to the LSGs in framing development projects. While providing infrastructure, a need-based approach may be followed, that too only after proper assessment of future requirements. Experience in decentralization shows that in local resource mobilization and improving efficiency, community participation can make significant contribution. In the Kerala context, this can be cherished by properly integrating district panchayat, village panchayat, schools and PTA on all matters pertaining to the functioning of schools. This type of integration may bring fruitful results in preventing dropouts, better evaluation, local resource mobilization, teacher training, skill development, preparation of study materials and promotion of various curricular, co-curricular and extra-curricular activities in schools.

NOTE

[Publication of the paper in 'Dimensions of Social Sector Development:Status, Challenges and Prospects' in G.Karath (ed) ISEC Monograph 8, 2005 is acknowledged].

REFERENCES

Ambasht, N.K (1998): "Elementary Education in Panchayathi Raj", in Aggarwal and Premi (eds) *Reforming School Education: Issues in Policy Planning and Implementation,* NIEPA, Vikas Publishing Pvt. Ltd. pp. 1-9.

Annigeri, V.B (1998): "Primary Education: How the Panchayats are Doing", in Aggarwal and Premi (eds.) *Reforming School Education: Issues in Policy Planning and Implementation,* NIEPA, Vikas Publishing Pvt. Ltd. pp. 40-48.

Archer, M (1979): *The Social Origins of Educational Systems*, London: Sage.

Boucher, L (1982): *Tradition and Change in Swedish Education London*: Pergamon.

Card, D and Krueger, A (1992): "Does School Quality Matter? Returns to Education and the Characteristics of Public Schools in United States", *Journal of Political Economy* 100(1): 1-40.

Carnoy, M (1993): "School Improvement: Is Privatization the Answer?" in J. Hannaway and M. Carnoy (eds), *Decentralization and School Improvement*, San Francisco: Jossey–Bass.

Chapman, Judit D., W. Boyd, R. Lander and D. Reynolds (1996): *The Reconstruction of Education: Quality, Equality and Control,* Chapman et. al (eds), London: Cassell.

Chubb, J and Moe, T (1990): *Politics, Markets and American Schools,* Washington D.C: Brooking Institution.

Clune, William H and John F. Witte (1990): "Choice and Control in American Education" in *The Theory of Choice and Control in American Education*, Vol. 1 Clune and Witte (eds), London and New York: Falmer Press.

Eskeland, Gunnar S and Filmer, Deon (2002): "Autonomy, Participation, and Learning in Argentine Schools: Findings and their Implications for Decentralization", Policy Research *Working Paper* 2766, The World Bank, Development Research Group, Public Services.

Gopinathan, S (1994): "Educational Development in a Strong Developmentalist State: The Singapore Experience", Paper presented to the Australian Association for Research in Education, Annual Conference.

Government of Kerala, State Planning Board (1997): Report of the Task Force on School Education, Ninth Five Year Plan, 1997–2002.

Government of Kerala, State Planning Board, Economic Review, 1997, 1998, 1999, 2000, 2001 and 2002.

Government of Kerala, State Planning Board (1998): Report of the Steering Committee on Education, Ninth Five Year Plan 1997–2002.

Government of Kerala, State Planning Board (2002): Planning by Local Self Governments: Tenth Five Year Plan 2002–07, Health Sector.

Government of Kerala, State Planning Board (2002): Planning by Local Self Governments: Tenth Five Year Plan 2002–07, Education Sector.

Government of Kerala (2002): Guidelines for the Preparation of Tenth Five Year Plan by Local Governments [G.O. [MS] No. 20/2002/Plg. Dt. 6.6.2002.

Grant, Carl A (1979): *Community Participation in Education*, Allyn and Bacon, Inc. London.

Green, Andy (1997): "Educational Achievement in Centralized and Decentralized Systems", in A.H. Halsey et al [ed], *Education: Culture, Economy and Society*, Oxford University Press.

Green, A. (1990): *Education and State Formation,* London: Macmillan.

Hannaway, Jane and Carnoy (1993): *Decentralization and School Improvement: Can we fulfill the Promise?* Hannaway and Carnoy (eds), Jossey–Brass Publishers, San Francisco.

Hanushek, Eric (1986): "Economics of Schooling: Production and Efficiency in Public Schools", *Journal of Economic Literature,* 24 [Sept]: 1141–1177.

Hanushek, Eric (1994): *Making Schools Work: Improving Performance and Controlling Costs*, Washington DC: Brookings Institution.

Hanushek, Eric (1995): "Interpreting Recent Research on Schooling in Developing Countries", *The World Bank Research Observer,* 10 (2): 227-46.

Jimenez, Emmanuel and Yasu, Sawada (1999): "Do Community Managed Schools Work? An Evaluation of El Salvador's EDUCO Program", *World Bank Economic Review*, 13 (3): 415-441.

Kaestle, C (1983): *Pillars of the Republic: Common Schools and American Society, 1780-1960*, Torento: Hill and Wang.

Mahajan, Baldev (1998): "Decentralised Management of School Education and Panchayati Raj Institutions", in Aggarwal and Premi (eds) *Reforming School Education: Issues in Policy, Planning and Implementation* , NIEPA, Vikas Publishing Pvt. Ltd, pp. 20-29.

Majumdar, M and A. Vaidyanathan (1994): "Access to Education in India: Retrospect and Prospect", *Journal of Educational Planning and* Administration, Vol. 7, No. 4, New Delhi.

Mehta, C (1995): *Education for All*, Kanishka Publishers, Delhi.

Panchamukhi P.R (2000): "Social Impacts of Economic Reforms in India: A Critical Appraisal", *Economic and Political Weekly,* Vol. XXXV, No. 10.

Pant, S.K and Pandey, Alok (2001): "Social Sector in Liberalised Era", in Ruddar Datt (ed) *Second Generation Economic Reforms in India*, Deep and Deep Publications Pvt. Ltd, New Delhi.

Pillai, G (1992): "Akshara Keralam, Total Literacy Programme in Kerala: A Case Study", *Journal of Educational Planning and Administration, 6 (3) New Delhi.*

Premi, Kusum K (1994): "Management of Education under Panchayati Raj: A Review of Literature", Paper Presented at the Seminar on Management of Education under Panchayati Raj, NIEPA, New Delhi.

Ranganathan, S., Sethi, Vinitha and Singh R.P (1998): Educational Reform and Development Strategy, vol. I , *Educational Reform and Planning Challenge*, Sandarbh Publishers, New Delhi.

Salam, M.A (2001): "Attention Towards Social Sector in India", in Ruddar Datt (ed), *Second Generation Economic Reforms in India*, Deep and Deep Publications Pvt. Ltd, New Delhi.

Sapra, C.L (1998): "Training Implications of Decentralised System of Educational Planning and Management", in Aggarwal Yash and Premi, K.K (eds.) *Reforming School Education: Issues in Policy Planning and Implementation,* National Institute of Educational Planning and Administration, Vikas Publishing Pvt. Ltd, pp. 96-104.

Schoppa, J (1991): *Education in Japan: A Case of Immobilist Politics,* London: Routledge.

Sexton, S (1987): *Our Schools: A Radical Policy*, London: Institute of Economic Affairs.

Sinha, Amarjeeth (1998): *Primary Schooling in India*, Vikas Publishing House.

Trivandrum Municipal Corporation (n.d): Evaluation of Ninth Five Year Plan and Approach to the Tenth Five Year Plan: Draft Document.

Trivandrum Municipal Corporation (2002): Development Seminar: Tenth Five Year Plan (2002-07), Draft Project Document.

Visaria, P.D., Gumber, A. and Visaria, L (1993): "Literacy and Primary Education in India: 1980-81 to 1999-91, Differentials and Determinants", *Journal of Educational Planning and Administration,* Vol. 7, No. 1, New Delhi.

Winkler, Donald R (1989): "Decentralisation in Education: An Economic Perspective", Policy Planning and Research *Working Papers,* Education and Employment, Population and Human Resources Department, The World Bank, WPS, 143.

Witte, John F (1990): "Choice and Control: An Analytical Overview", in W. Clune and J. Witte (eds) *Choice the theory of and Control in American Education*, Vol. I, New York: Falmer Press.

CONSOLIDATED BIBLIOGRAPHY

Abed, George T and Gupta, Sanjeev (2002): "The Economics of Corruption: An Overview", in *Governance, Corruption and Economic Performance*, [eds] George T. Abed and Sanjeev Gupta, IMF, Washington D.C, pp.1-16.

A Handbook of Travancore, Kerala Vilasom Press, Trivandrum (1881): Vol. I, p. 327.

Ahlin, Christian (2004): "Corruption and Anti-corruption Mechanisms under Decentralization", *JEL* No. D 73, h 77, O17,1.

A Judicial History of Travancore (1932): Published on the occasion of the Golden Jubilee of the High Court, Appendix I, High Court, p. iii.

Alsop, R., Krishna, A and Sjoblom, D (2000): "Are Grama Panchayats Inclusive?" Report of a study conducted in Rajasthan and MP, Background Paper No. 3, World Bank unpublished project work, *Overview of Rural Decentralization in India*, Vol. 3.

Ambasht, N.K (1998): "Elementary Education in Panchayathi Raj", in Aggarwal and Premi (eds) *Reforming School Education: Issues in Policy Planning and Implementation,* NIEPA, Vikas Publishing Pvt. Ltd. pp. 1-9.

Annigeri, V.B (1998): "Primary Education: How the Panchayats are Doing", in Aggarwal and Premi (eds.) *Reforming School Education: Issues in Policy Planning and Implementation,* NIEPA, Vikas Publishing Pvt. Ltd, pp. 40-48.

Archer, M (1979): *The Social Origins of Educational Systems*, London: Sage.

Arkadie, Brain Van (1989): "The Role of Institutions in Development", *Proceedings of the World Bank Annual Conference on Development Economics,* The World Bank, 1990, pp. 153-179.

Bardhan, Pranab (1989): "Comment on the Role of Institutions in Development by Van Arkadie", *Proceedings of the World Bank Annual Conference on Development Economics,* The World Bank, 1990, pp. 181-185.

Bardhan, Pranab and Mookherjee, Dilip (2002): Corruption and Decentralization of Infrastructure Delivery in Developing Countries, forming Part 2 of their *Working Paper* "Expenditure Decentralization and Delivery of Public Services in Developing Countries".

Bardhan, Pranab and Mookherjee, Dilip (2005): 'Decentralization, Corruption and Government Accountability: *An Overview* 'For *'Handbook of Economic Corruption'* [ed] Susan Rose- Ackerman, Edward Elgar.

Baxi, Upendra (1980): *The Crisis of the Indian Legal System*, Vikas Publishing House, p. 324.

Baxi, Upendra (1986): "Participatory Justice: A Critique of Nyaya Panchayats in Karnataka", in George Mathew (ed) *Panchayati Raj in Karnataka Today: Its National Dimensions*, Institute of Social Sciences and Concept Publishing Co. New Delhi, pp. 76-87.

Beier, Christoph (1999): "The Impact of Decentralization on Rural Development", *Agriculture and Rural Development,* Vol. 6, No. 1, pp. 18-21.

Bhargava, Vinay and Bolongaita, Emil (2004): *Challenging Corruption in Asia: Case Studies and a Framework for Action*, The World Bank, Directions in Development, Washington D.C.

Bird, Richard (1994): "Financing Local Services: Patterns, Problems and Possibilities", Paper prepared for *Global Report on Human Settlements,* April.

Boucher, L (1982): *Tradition and Change in Swedish Education,* London: Pergamon.

Brett, E.A (1993): "Voluntary Agencies as Development Organizations: Theorizing the Problem of Efficiency and Accountability", *Development and Change*, 24 [2] pp. 269–303.

Campbell, T., Peterson, G and Brakarz, J (1991): "Decentralization to Local Government in LAC: National Strategies and Local Response in Planning, Spending and Management", Latin America and the Caribbean Technical Departments, *Regional Studies Programme,* Report No. 5, Washington D.C.: World Bank.

Card, D and Krueger, A (1992): "Does School Quality Matter? Returns to Education and the Characteristics of Public Schools in United States", *Journal of Political Economy* 100 (1): 1-40.

Carnoy, M (1993): "School Improvement: Is Privatization the Answer?" in J. Hannaway and M. Carnoy (eds), *Decentralization and School Improvement*, San Francisco: Jossey--Bass.

CESS et. al (1991): Panchayat Level Resource Mapping: An Approach Paper- A Model for Micro Level Resource Survey with People's Participation.

Chapman, Judit D., W. Boyd, R. Lander and D. Reynolds (1996): *The Reconstruction of Education: Quality, Equality and Control,* Chapman et. al (eds), London: Cassell.

Cheema, G.S and Rondinelli, D.A (1983): *Decentralization and Development: Policy Implementation in Developing Countries*, Cheema and Rondinelli (eds), Beverly Hills: Sage.

Chubb, J and Moe, T (1990): *Politics, Markets and American Schools,* Washington D.C: Brooking Institution.

Clune, William H and John F. Witte (1990): "Choice and Control in American Education" in *The Theory of Choice and Control in American Education*, Vol. 1, Clune and Witte (eds), London and New York: Falmer Press.

Cochin Information (1944): Vol. III, No. 2, Issued by the War Publicity Department, February-March, p. 10.

Cochin Information (1945): Vol. IV, No. 4, April-June, p. 20.

Conyers, D (1990): "Decentralization and Development Planning: A Comparative Perspective", in P. de Valk and K.H. Wekwete [eds] *Decentralizing for Participatory Planning,* Aldershot, Avebury.

Cooke, P (1990): *Back to the Future*, Unwin Hyman, London.

Cooke, P (1995): *Reinventing the Region: Firms Clusters and Networks,* Friedrich Ebert Stiftung, Bonn.

Cranston, Maurice (1973): *What are Human Rights?*, The Bodley Head London, p. l.

Development Report (1997-2001): Grama Panchayats in Thiruvananthapuram District.

Doig, Alan and Theobald, Robin (2000): "Why Corruption?", in *Corruption and Democratization*, [eds] Alan and Robin, Frank Cass, London, p. 5.

Douglas, James (1983): *Why Charity? The Case for a Third Sector,* Sage Publications, Beverly Hills.

Dubey, Ajay (1995): "Strengthening Panchayati Raj Institutions: Strategies for NGOs", *Working Paper* 87, Institute of Rural Management, Anand, pp. 1-10.

ESCAP, UN (1991): "Fiscal Decentralization and the Mobilization and Use of National Resources for Development: Issues, Experience and Policies in the ESCAP Region", *Development Papers* 11, pp. 1-37.

Eskeland, Gunnar S and Filmer, Deon (2002): "Autonomy, Participation, and Learning in Argentine Schools: Findings and Their Implications for Decentralization", Policy Research *Working Paper* 2766, The World Bank, Development Research Group, Public Services.

Ettlinger, N (1994): "The Localisation of Development in Comparative Perspective," *Economic Geography,* 70(2): pp. 1444-66.

Fisman, R and Gatti, R (2002 a): "Decentralization and Corruption Across Countries", *Journal of Public Economics*, 83, 325-345.

Fisman, R and Gatti, R (2002 b): "Decentralization and Corruption: Evidence from US Federal Trånsfer Programs", *Public Choices*, 113, 25-35.

Frances and Bee, Ronald (1998): *Facilitation Skills*, Institute of Personnel and Development, United Kingdom, Appendix-D: Force–Field Analysis.

Friedrich, Carl J (1966): "Political Pathology", *Political Quarterly*, 37: p. 74.

Giridharan, M.P (2001): "Involvement of Women in Planning, Implementation and Monitoring of Projects Under People's Campaign for Decentralized Planning", *Evaluation Studies on Decentralized Planning*, State Planning Board, GOK.

Gopinathan, S (1994): "Educational Development in a Strong Developmentalist State: The Singapore Experience", Paper presented to the Australian Association for Research in Education, Annual Conference.

Government of India, Law Commission (1958): *Reform of Judicial Administration*, Fourteenth Report.

Government of India (1973): *Report of the Expert Committee on Legal Aid : Processual Justice to the People*, Chairman Justice V.R. Krishna Iyer.

Government of India (1977): *Report on National Judicare: Equal Justice-Social Justice*, Chairman, Justice Bhagawathi.

GOI, Planning Commission (1991): *Report of the Committee on Model Cooperatives Act,* Government of India, New Delhi, pp. 1-57.

Government of India (2000): *Report of the Eleventh Finance Commission*, New Delhi.

Government of Kerala (1979): Committee for Artisan Development, Project Report.

Government of Kerala, State Planning Board (1991): *Report of the Steering Committee on Decentralized Planning and Planning for Rural Development*, Eighth Five Year Plan 1990-95.

Government of Kerala, Department of Economics and Statistics (1995): Report of Wage Structure 1982-92.

Government of Kerala, Department of Economics and Statistics (1996): Panchayath Level Statistics, Thiruvananthapuram District.

Government of Kerala (1996 a): Ninth Five Year Plan–Decentralization and People's Participation–Mass Campaign, GO. (MS) No. 10/96/plg Dated 30th July 1996, Planning & Economic Affairs (A) Department.

Government of Kerala, State Planning Board (1996 b): *People's Campaign for 9th Plan: An Approach Paper.*

Government of Kerala (1996c): Ninth Five Year Plan–People's Campaign for Formulation of Plan–High Level Guidance Council–constituted, GO (MS) No. 12/96 plg dt 8th August.

Government of Kerala, State Planning Board (1997): Report of the Task Force on School Education, Ninth Five Year Plan, 1997–2002.

Government of Kerala, State Planning Board, Economic Review, 1997, 1998, 1999, 2000, 2001 and 2002.

Government of Kerala, State Planning Board (1998): Report of the Steering Committee on Education, Ninth Five Year Plan 1997–2002.

Government of Kerala (1998): Report of the Committee on Decentralization of Powers (S.B. Sen, Chairman).

Government of Kerala, State Planning Board (1999): *Gunabhokthrusamithy Pravarthana Margarekha* (Malayalam).

Government of Kerala (2000): *"People Friendliness-Transparency: New Perspectives"* (Malayalam), Papers presented in the International Conference on Democratic Decentralization (ICDD), Trivandrum, May 23-27, [eds] Jagajeevan et. al.

Government of Kerala, State Planning Board (2001): *Evaluation Studies on Decentralized Planning*: Summary of Reports.

Government of Kerala, State Planning Board (2002): Planning by Local Self Governments: Tenth Five Year Plan 2002–07, Health Sector.

Government of Kerala, State Planning Board (2002): Planning by Local Self Governments: Tenth Five Year Plan 2002–07, Education Sector.

Government of Kerala (2002): Guidelines for the Preparation of Tenth Five Year Plan by Local Governments [G.O. [MS] No. 20/2002/Plg. Dt. 6.6.2002.

Government of Kerala (2003): "Modernizing Government Programme [MGP], Theme V, Effective, Efficient and Accessible Local Self Government", Schematic Outlines of Detailed Implementation Plan, pp. 34-35.

Government of Kerala, State Planning Board (2005): Economic Review 2004.

Grant, Carl A (1979): *Community Participation in Education*, Allyn and Bacon, Inc. London.

Gray, Chery W., Lynn S. Khadiagala and Richard J. Moore (1990): "Institutional Development Work in the Bank-A Review of 84 Bank Projects", Policy, Research and External Affairs *Working Paper,* Development Economics, WPS 437, The World Bank, Washington D.C., pp. 1-27.

Green, A (1990): *Education and State Formation.* London: Macmillan.

Green, Andy (1997): "Educational Achievement in Centralized and Decentralized Systems", in A.H. Halsey et.al [ed], *Education: Culture, Economy and Society*, Oxford University Press.

Gupta, I.D (2002): "Corruption and Rent Seeking in Les Developed Countries: An Indian Perspective", *Indian Journal of Economics*, No. 328, July, Vol. LXXXIII, pp. 75-93.

Hannaway, Jane and Carnoy (1993): *Decentralization and School Improvement: Can we Fulfill the Promise?,* Hannaway and Carnoy (eds), Jossey–Brass Publishers, San Francisco.

Hanushek, Eric (1986): "Economics of Schooling: Production and Efficiency in Public Schools", *Journal of Economic Literature,* 24 [Sept]: 1141–1177.

Hanushek, Eric (1994): *Making Schools Work: Improving Performance and Controlling Costs*, Washington DC: Brookings Institution.

Hanushek, Eric (1995): "Interpreting Recent Research on Schooling in Developing Countries", *The World Bank Research Observer* 10 (2): 227-46.

Haragopal, G (1994): *"Panchayati Raj and Human Rights"*, Occasional Papers-6, Renewing Local Self Government in Rural India, Govt. of India, Ministry of Rural Development, Department of Rural Development, NIRD, Hyderabad.

Haragopal, G (1997): *Political Economy of Human Rights: Emerging Dimensions*, Himalaya Publishing House.

Harilal, K.N (2005): "New Vistas of Local Governance: A Study of Democratic Decentralization in Kerala", Paper presented in the International Conference on 'A Decade of Decentralization in Kerala: Issues, Options and Lessons', Organised by Institute of Social Sciences, New Delhi, in Thiruvananthapuram, 7-9, October.

Heidenheimer, Arnold J., Michael, Johnston and Victor, T.Le Vine (1989): *Political Corruption: A Handbook*, (eds) Heidenheimer et. al, New Brunswick, NJ: Transaction Books.

Herlaar, Martin and Wim Olthof (1994): "Rural Local Government and Development Planning, with Experiences from Buhera District, Zimbabwe", *Working Paper* Series No. 181, Institute of Social Studies, The Netherlands, pp. 1-42.

Hunter, Guy (1969): *"Modernising Peasant Societies",* Oxford University Press, New York.

Huther, Jeff and Shah, Anwar (1998): "Applying a Simple Measure of Good Governance to the Debate on Fiscal Decentralization", World Bank Policy Research *Working Paper* 1894, Washington D.C.

Hussi, Pekka Josette Murphy, Ole, Lindberg and Lyle, Brenneman (1993): *"The Development of Cooperatives and other Rural Organizations-The Role of the World Bank",* World Bank Technical Paper No. 199, Africa Technical Department Series, The World Bank, Washington D.C, pp. 1-94.

ILO (1990): *"The Socio-Economic Impact of Technical Cooperation Projects Concerning Rural Development",* ACRD/XI/1990//111, International Labour Office, Advisory Committee on Rural Development, Geneva.

ILO (1992): "*Democratization and the ILO",* International Labour Conference, 79th Session, Report of the Director General, Geneva.

Innes, C.A (n.d): Malabar (Gazetteer), FD Evans (ed) Vol. I and II Reprinted by the State Editor, Kerala Gazetteers, Thiruvananthapuram, 1997, p. 385.

Institute of Social Sciences (2001): Panchayati Raj Update, Vol. 11, No. 10, 94.

Institute of Social Sciences (2001): *Panchayati Raj Update*, Vol. VIII, No. 12, 96, December, p. 3, 8.

Institute of Social Sciences (2002): *Panchayati Raj Update*, Vol. IX, No. 9, 105, Sept, p. 5.

Institute of Social Sciences (2002): "Decentralisation as an Instrument of Human Rights Promotion–I", Edited excerpts from 'Local Rule: Centralization and Human Rights'- published by International Council on Human Rights Policy, Geneva, *Panchayati Raj Update,* Vol. IX, No.10, October, p. 6.

Institute of Social Sciences (2002): "Decentralisation as an Instrument of Human Rights Promotion – II", Edited excerpts from 'Local Rule: Centralization and Human Rights'- published by International Council on Human Rights Policy, Geneva, *Panchayati Raj Update,* Vol. IX, No. 11, 107, November, p. 6-7.

Institute of Social Sciences (2002): "Decentralisation as an Instrument of Human Rights Promotion–III", Edited excerpts from 'Local Rule: Centralization and Human Rights'- published by International Council on Human Rights Policy, Geneva, *Panchayati Raj Update,* Vol. IX, No. 12, 108, December, p. 6.

Institute of Social Sciences (2004 & 2005): '*Panchayati Raj Update*', Various Issues.

International Conference on Democratic Decentralisation (2000): *Transparency: New Perspectives* (Malayalam), N. Jagajeevan et. al [ed.] collection of papers, p. 250-260, 304.

Israel, Arturo (1987): *Institutional Development-Incentives to Performance,* The World Bank, Washington D.C, The Johns Hopkins University Press.

Issac, Thomas T.M and Harilal K.N (1997): "Planning for Empowerment: People's Campaign for Decentralized Planning in Kerala", *Economic and Political Weekly*, Volume XXXII, Nos. 1 and 2, January 4–11, pp. 53-58.

Issac, Thomas T.M (1999): "People's Planning and NHGs; Experience and Lessons" (Malayalam), in T.N. Seema et.al (ed) Ayalkootta Sangamom, State Planning Board, p. xi.

Iyer, Krishna V.R (1979): *Of Law and Life*, Vikas Publishing House, p. 141.

Jacob, Alice (n.d): "Rural Litigation: Role of Nyaya Panchayats as Agents of Grassroots Justice", Appendix II in S.N. Mathur (1997), *Nyaya Panchayats as Instruments of Justice*, ISS, New Delhi.

Jaswal, Paramjit S (1996): *Directive Principles: Jurisprudence and Socio-Economic Justice in India*, APH Publishing Corporation, New Delhi.

Jha, S (2000): "Fiscal Decentralization in India: Strengths, Limitations and Prospects for PRIs", Background Paper No. 2, World Bank, unpublished project work, *Overview of Rural Decentralization in India*, Vol.3.

Jimenez, Emmanuel and Yasu, Sawada (1999): "Do Community Managed Schools Work?: An Evaluation of El Salvador's EDUCO Program", *World Bank Economic Review*, 13 (3): 415-441.

John, M.S and Chathukulam, Jos (2001): "Social Capital Formation and Institutional Growth Following Participatory Planning in Kerala: A Critical Appraisal", in *Emerging Institutions for Decentralized Rural Development*, [ed] S.P. Jain, Proceedings of NIRD Foundation Day Seminar, January 7-8, Vol. I, NIRD Hyderabad, pp. 478-489.

Kaestle, C (1983): *Pillars of the Republic: Common Schools and American Society, 1780-1960*, Torento: Hill and Wang.

Kaushik, P.D (2005): "Panchayati Raj Movement in India: Retrospective and Present Status", in Bibek Debroy and P.D. Kaushik (eds) *Energizing Rural Development Through Panchayats*, Academic Foundation and RGICS.

Kerala State Planning Board (1996): *Report of the Committee on Reform of Cooperatives in Kerala,* pp. 1-25.

Kerala State Planning Board (1997): The People's Campaign for Ninth Plan in Kerala: An Introduction.

Koenig, Shulamith (1994): "Defending the Future", in Reoch Richard (ed) *Human Rights: The New Consensus*, Regency Press in association with the UNHCR, p. 45, 46.

Kohli, Suresh (1975): *Corruption in India*, New Delhi, Chetan Publications.

Korten, David (1980): "Community Organization and Rural Development: A Learning Process Approach, *Public Administration Review,* Vol. 40, No. 5, pp. 480-511.

Koshy, Ninan (2000): "Human Rights under Globalisation", *South Asian Journal of Socio Political Studies*, Vol. 1, No. 2, June-December, pp. 14-20.

Krishnakumar, R (2005): "Power to the People Eroded", *The Hindu Daily*, January 2, p. 14.

Krishna, Sudhir (1994): "Restructuring the Institutional Arrangements to Strengthen the Panchayati Raj," in Amitava Mukherjee (ed) *Decentralisation: Panchayat in the Nineties,* Vikas Publishing House, New Delhi, pp. 189-201.

KRPLLD, Center for Development Studies (1995): Kalliasseri Experiment in Local Level Planning, Seminar on Local Level Planning-Learning from Kalliasseri, 2-3, October.

KRPLLD, Center for Development Studies (2003): Annual Report 6 (2001-02): Panchayat Level Development Planning: Problems and Prospects of People's Participation in a Party-dominated Democracy, pp. 27-44.

Kumar, Somesh (1999): Applications of Force Field Analysis in PRA, *Exchanges*, Issue Nos. 25 & 26, June and September.

Kudumbasree, Thiruvananthapuram District (1999): Community Development Society Handbook, [Malayalam].

Kushawaha, R (1977): *Working of Nyaya Panchayats in India: A Case Study of Varanasi District*, Young Asia Publications, New Delhi, p. 105.

Levi, Y and H Litwin (1986): *Community and Cooperatives in Participatory Development,* Gower London.

Lewin, Kurt (1951): *Field Theory in Social Science,* Harper and Row, New York.

Leys, Colin (1989): "What is the Problem about Corruption", in *Political Corruption: A Handbook,* (eds) Arnold J. Heidenheimer., Michael Johnston, and Victor T, Le Vine, New Brunswick, NJ: Transaction Books, pp. 51-66.

Linstone, H.A and M. Turoff (1975): *The Delphi Method: Techniques and Applications,* Mass: Addision Wesley, Reading.

Menon, Madhava N.R (2000): Sri Chithira Thirunal Memorial Lecture, University of Kerala, March.

Menon, Madhava N.R (2000 a): "Social and Economic Rights: Poverty is Greatest Threat", in Mathews, George Chunakara (ed) *Human Rights 2000*, Institute of Human Rights, Vigil India Movement, Bangalore, pp. 5-12.

Maetz, Materne and Maria, G. Quieti (1987): *Training for Decentralized Planning: Lessons from Experience,* Vol.1, Development Policy Studies and Training Service, Policy Analysis Division, FAO of the UN, Rome, pp. 1-61.

Mahajan, Baldev (1998): "Decentralised Management of School Education and Panchayati Raj Institutions", in Aggarwal and Premi (eds) *Reforming School Education: Issues in Policy, Planning and Implementation* , NIEPA, Vikas Publishing Pvt. Ltd, pp. 20-29.

Majumdar, M and A. Vaidyanathan (1994): "Access to Education in India: Retrospect and Prospect", *Journal of Educational Planning and* Administration, Vol. 7, No. 4, New Delhi.

Mancur, Olson (1996): "Collective Choice Perspectives on Indian Economy", Redistribution as Means of Economic Reforms Mooted. *Financial Express, January 13.*

Mathew, Anand (2005): "Resource Mobilization and Intergovernmental Transfers", Paper presented in the International Conference on 'A Decade of Decentralization in Kerala: Issues, Options and Lessons', Organised by Institute of Social Sciences, New Delhi, in Thiruvananthapuram, 7-9, October.

Mathew, George (1995): A Tale of Two Panchayats. *The Hindu* Daily, July 6.

Mathew, George (2001): "Panchayat Elections: Dismal Record", *Economic and Political Weekly*, 20 January.

Mathew, George (2003): "Panchayati Raj Institutions and Human Rights in India", *Economic and Political Weekly*, Vol. XXXVIII, No. 2, pp. 155-162.

Mathur, S.N (1997): *Nyaya Panchayats as Instruments of Justice*, Institute of Social Sciences, Concept Publishing Company, New Delhi.

Mauro, Paolo (1995): "Corruption and Growth", *Quarterly Journal of Economics,* Vol. 10, August, pp. 681-712.

Mauro, Paolo (1997a): "Why Worry About Corruption?", *Economic Issues –6*, International Monetary Fund, Washington D.C, p. 1.

Mauro, Paolo (1997b): "The Effects of Corruption on Growth, Investments, and Government Expenditure: A Cross Country Analysis", in *Corruption in the Global Economy*, [ed] Kimberly Ann Elliott, Institute for International Economics, Washington.

Mawhood, P. (1983): *Local Government in the Third World–The Experience of Tropical Africa,* Mawhood, P. (ed.), John Wiley and Sons, Chickester.

McLean, Keith (2000): "Political Decentralisation: International Experience", paper presented in the IMG Seminar, Thiruvananthapuram, Kerala.

McLean, Keith (2001): "Political Decentralization: International Experience", Paper presented in the International Conference on Democratic Decentralization in Kerala: Status and Next Steps, Trivandrum, August 17-18.

Mehta, C (1995): *Education for All*, Kanishka Publishers, Delhi.

Mello, Luis de and Barenstein, Matias (2002): "Fiscal Decentralization and Governance: A Cross Country Analysis", in *Governance, Corruption and Economic Performance*, [eds] George T. Aded and Sanjeev Gupta, IMF, Washington D.C, pp. 333-365.

Minis, Henry and Dennis, Rondinelli (1989): "Promoting Economic Development and Employment Generation Through Decentralization in Senegal", USAID *Working Paper,* Mimeo, p. 122.

Mohanan, B (2000): "Controlling Corruption at the Grossroots: Lessons from a Decentralized and Participatory Development Administration", in *Corruption at the Grass roots: The Shades and Shadows* [eds] Narayanaswamy N., M.P. Boraian and M.A. Jayaraju, Concept Publishing Company, New Delhi, pp. 116-136.

Mohanan, S (2001): "Resource Mobilization in People's Planning: A Case Study", *Evaluation Studies on Decentralized Planning*, State Planning Board, GOK.

Moody–Stuart G (1994): *Grand Corruption in Third World Development*, Transparency International, Berlin.

Mukhopadyay, Amitabh (2005): Accountability and Communicative Action: An Alternative Imaginary, *Economic and Political Weekly*, Vol. XL. No. 52, pp. 5463-69.

Muralikrishna, M (2001): "Evaluation of Panchayat Level Resource Mapping for Panchayat Level Planning", *Evaluation Studies on Decentralized Planning*, State Planning Board, GOK.

Narayana, D (2005): "Local Governance Without Capacity Building: Ten Years of Panchayati Raj", *Economic and Political Weekly,* June 25, pp. 2822-32.

Nathan, Iben et. al (1994): "Brief Guidelines for the Institutional Screening of Local Governments", CDR Local Institutions and Rural Development Group, CDR *Working Paper* 94.5, Denmark pp. 1-9.

National Productivity Council (1979): *Rural Artisans in Payyannur: Report of a Study Sponsored by the Inter Agency Co–ordination Committee for Decentralized Sector.*

Nugent, Jeffrey B (1993): "Between State, Markets and Households: A Neoinstitutional Analysis of Local Organisations and Institutions", *World Development,* 21 (4), pp. 623-632.

Nye, Joseph S (1967): "Corruption in Political Development: A Cost- Benefit Analysis", *American Political Science Review*, 61, June, p. 417.

Olowu, D (1988): *African Local Governments as Instruments of Economic and Social Development,* International Union of Local Authorities, The Hague.

Oommen, M.A (2004): "Decentralization and Poverty Reduction", UNDP Study, as Reported in Economic Review 2004, Government of Kerala, p. 466.

Oommen, M.A (2004): Preface to "Fiscal Decentralization in Kerala", in Geeta Sethi (ed) *Fiscal Decentralization to Rural Governments in India*, The World Bank, Oxford University Press.

Pal, Mahi (1999): "Functioning of the National Human Rights Commission–A Brief Review", in *State of Human Rights in India 1999*, Sponsored Publication, Indian Social Institute, New Delhi, p. 7.

Pal, Mahi (2000): "Panchayats and Human Rights", *Kurukshetra*, Vol. 48, No.11, August, pp. 32-36.

Panchamukhi P.R (2000): "Social Impacts of Economic Reforms in India: A Critical Appraisal", *Economic and Political Weekly,* Vol. XXXV. No. 10.

Pant, S.K and Pandey, Alok (2001): "Social Sector in Liberalised Era", in Ruddar Datt (ed.) *Second Generation Economic Reforms in India*, Deep and Deep Publications Pvt. Ltd, New Delhi.

Pareek, Udai (1981): *Beyond Management*, New Delhi, Oxford and IBH.

Parker, Andrew N (1995): "Decentralisation: The Way Forward for Rural Development", *Policy Research Working Paper* 1475, Agriculture and Natural Resources Department, Sector Policy and Water Resources Division, The World Bank, p. 23.

Pavarala, Vinod (1996): *Interpreting Corruption: Elite Perspectives in India*, Sage Publications, New Delhi.

Pillai, G (1992): "Akshara Keralam, Total Literacy Programme in Kerala: A Case Study", *Journal of Educational Planning and Administration,* 6 (3) New Delhi.

Pillai, P.P (2005): "Capacity Building for Decentralized Democratic Governance: Some Lessons from Kerala Experience", Paper presented in the International Conference on ' A Decade of Decentralization in Kerala: Issues, Options and Lessons', Organized by Institute of Social Sciences, New Delhi, in Thiruvananthapuram, 7-9, October.

Prakash, B.A (2005): "Plan Performance of LSGs: A Study in 36 Grama Panchayats in Kerala", Paper presented in the International Conference on 'A Decade of Decentralization in Kerala: Issues, Options and Lessons', Organised by Institute of Social Sciences,New Delhi, in Thiruvananthapuram, 7-9, October.

Premi, Kusum K (1994): "Management of Education under Panchayati Raj: A Review of Literature", Paper presented at the Seminar on Management of Education under Panchayati Raj, NIEPA, New Delhi.

Prudhomme, Remy (1994): "One the Dangers of Decentralisation", Policy Research *Working Paper* 1252, Transportation, Water and Urban Development Department, The World Bank, pp. 1-35.

Putnam, R (1993): *Making Democracy Wo*rk, Princeton: Princeton University Press.

Raghaviah, Maleeha (2002): *Jagratha Samithies Hold the Key*, The Hindu Daily, July, 29.

Rajan, J.B (2002): "Declining Indigenous Regulations on Fishery: A Case of Kadakkodi (Sea Court) in Kerala, South India", Fisheries Research Cell, Programme for Community Organisation (PCO), Thiruvananthapuram (paper presented in the XIIIth International Congress on Legal Pluralism and Unofficial Law in Social, Economic and Political Development", organised by the Commission on Folk Law and Legal Pluralism, Nepal, Chiang Mai, Thailand, 7-10 April.

Rajaraman, Indira., O.P Bohra and V.S Renganathan (1996): "Augmentation of Panchayat Resources", *Economic and Political Weekly*, XXXI(18), May 4, pp. 1071-1083.

Rajasekharan, K (2004): "Citizens' Charter: A New Tool for Vitalizing Local Government Services in Kerala", *Panchayati Raj Update*, Vol. XI, No.10, 130, October, p. 8.

Ramachandran, V (1993): *State Legislation on Panchayati Raj Consequent on the Constitution (Seventy-third Amendment) Act, 1992, Issues for Discussion,* State Planning Board, Thiruvananthapuram, Kerala, pp. 1-29.

Ranganathan, S., Sethi, Vinitha and Singh R.P (1998): Educational Reform and Development Strategy, Vol. I , *Educational Reform and Planning Challenge*, Sandarbh Publishers, New Delhi.

Reoch, Richard (1994): *Human Rights: The New Consensus*, Regency Press in Association with the UNHCR, p. 45, 46.

Reserve Bank of India (1974): All India Debt and Investment Survey 1971–72.

Rondinelli, D.A (1981): "Government Decentralization in Comparative Perspective: Theory and Practice in Developing Countries", *International Review of Administrative Science*, 47.

Rondinelli, D.A (1983): "Decentralization of Development Administration in East Africa", in Cheema G.S and Rondinelli D.A (eds) *Decentralisation and Development: Policy Implementation in Developing Countries*, Sage, Beverly Hills, pp.77-125.

Rondinelli, Dennis and Nellis, John (1986): "Assessing Decentralization Policies in Developing Countries: The Case for Cautious Optimism", *Development Policy Review*, 4:3-23.

Sahay, Gaurang R (2005): "Decentralization, Panchayati Raj and the Rights of Marginalized Groups in India: A Case Study for Civil Society", paper presented in the NIRD Foundation Day Seminar 9-10, November 2005, National Institute of Rural Development, Hyderabad.

Salam, M.A (2001): "Attention Towards Social Sector in India", in Ruddar Datt (ed), *Second Generation Economic Reforms in India*, Deep and Deep Publications Pvt. Ltd, New Delhi.

Sapra, C.L (1998): "Training Implications of Decentralised System of Educational Planning and Management", in Aggarwal Yash and Premi, K.K (eds) *Reforming School Education: Issues in Policy Planning and Implementation*, National Institute of Educational Planning and Administration, Vikas Publishing Pvt. Ltd, pp. 96-104.

Sankaran, P.N (1994): "Excess Liquidity Syndrome in the Co-operative Banking System of Kerala", *International Congress on Kerala Studies, Abstracts*, Vol. 5, AKG Centre for Research and Studies, Thiruvananthapuram, pp. 47-48.

Sankaran, P.N (1996): "Institutional Reforms for Panchayati Raj: Towards Linking Cooperatives with Local Governments", in R. Rajagopalan [ed.] *Rediscovering Cooperation*, Volume. I, Bases of Cooperation, Institute of Rural Management, Anand, Gujarat.

Sankaran, P.N (1997): "Panchayath Raj: Institutional Dimensions and Deficiencies", *Kurukshetra*, Volume XLV, No. 9 June.

Sankaran, P.N (1998): "Decentralized Planning: Opportunities for NGOs", *Employment News*, Volume XXIII, No. 28 10 -16 October.

Sankaran, P.N (1999): "Empowerment and Welfare of Artisans: The Kerala Experience", *Kurukshetra*, Volume 48, No. 2 November.

Sankaran, P.N and Vijayakumar, B (2001): "Proliferation and Interface of Institutions Under Decentralized Planning: The Kerala Experience", in *Emerging Institutions for Decentralized Rural Development*, [ed] S.P Jain, Proceedings of NIRD Foundation Day Seminar, January 7–8, Vol. 1, NIRD Hyderabad, pp. 446–464.

Sankaran, P.N (2001): "Towards an Artisan Plan", *Evaluation Studies on Decentralized Planning*, State Planning Board, Government of Kerala.

Sankaran, P.N (2005): "Towards Transparency and Corruption-free Governance: An Overview of Decentralization in Kerala", in A. Ranga Reddy [ed.] *Dimensions of Crime and Corruption in India*, Serials Publications, pp. 160-185.

Schonwalder, Gerd (1992): "New Ways of Linking The State and Civil Society: Experiences with Local Government in Lima, Peru", *CDAS Discussion Paper* No. 766, pp. 1-19.

Schoppa, J (1991): *Education in Japan: A Case of Immobilist Politics*, London: Routledge.

Shah, Anwar and Thompson, Theresa (2004): "Implementing Decentralised Local Governance: A Treacherous Road with patholes, detours and road closures", *Panchayat Raj Update*, Vol. X1, No. 1, 121, January, pp. 6-7.

Sheth, P (1991): "Panchayati Raj Revival: Implications and Imperatives", in Hooja and Mathur (ed) *District and Decentralized Planning*.

Sengenberger, W (1993): "Local Development and International Economic Cooperation", *International Labour Review*, 132 (3) 315.

Sexton, S (1987): *Our Schools: A Radical Policy*, London: Institute of Economic Affairs.

Singvi, L.M (1977): Foreword to *Working of Nyaya Panchayats in India: A Case Study of Varanasi District*, Kushawaha R, The Institute of Constitutional and Parliamentary Studies, Young Asia Publications, New Delhi, p. ii and iii.

Sinha, Amarjeeth (1998): *Primary Schooling in India*, Vikas Publishing House.

Socio Economic Unit Foundation, Kerala (2001): "Workshop on Construction Works under the People's Planning Campaign, Kerala: Social and Community Contracting Issues", *Background Papers*, 24 May, Trivandrum.

Sreekumar, G.S and Gopakumar, V (2005): 'Citizens' Charter: A Tool for Social Auditing', Paper presented in the International Conference on ' A Decade of Decentralization in Kerala: Issues, Options and Lessons' organized by the Institute of Social Sciences, 7-9 October, Thiruvananthapuram, Kerala.

Streetan, Paul (1991): "Markets and States", Paper prepared for a Conference on *The State and Market in Development*, Brown University, Providence, RI.

Stockmayer, Albrecht (1999): "Decentralization: Global Fad or Recipe for Sustainable Local Development?", *Agriculture and Rural Development*, Vol. 6, No. 1, April, Chapter on Decentralization, p. 3.

Sudhakaran, M.N (2004): "Plan Integration in the Context of Multi level Planning – A Study based on Thrissur District", *Discussion Paper* No.102, Kerala Research Programme on Local Level Development, Center for Development Studies, Kerala.

Taimni K.K (1993): *Cooperative Development: The Next Phase*, ICA Domus Trust, New Delhi.

Tanzi, Vito (1995): "Fiscal Federalism and Decentralization: A Review of Some Efficiency and Marco Economic Aspects", Paper presented at the *Annual Bank Conference on Development Economics*, Washington D.C. May 1–2.

Tanzi, Vito and Hamid, Davoodi (1997): "Corruption, Public Investment and Growth", *IMF Working Paper* 97/139, Washington, IMF.

Tanzi, Vito and Hamid, Davoodi (1998): "Roads to Nowhere: How Corruption in Public Investment Hurts Growth", *Economic Issues–12*, International Monetary Fund, Washington D.C. p. 10.

Tanzi, Vito (1998): "Corruption Around the World: Causes, Consequences, Scope and Cures", *IMF Staff Papers*, Vol. 45, December, pp. 559–594.

The Economist (1991): *"A Survey of India"*, May 4, pp. 3-18.

The World Bank (2002): "Decentralization in Kerala", Chapter III, *Rural Decentralization in India*.

The World Bank (2004): Fiscal Decentralization to Rural Governments in India, [ed] Geeta Sethi, Oxford University Press.

Thomas, Baby (2001): "The Impact of Participatory Decentralized Planning on the Nature of Projects and Own Resource Generation of Panchayats in Kerala", *Evaluation Studies on Decentralized Planning,* State Planning Board, GOK.

Trivandrum Municipal Corporation (n.d): Evaluation of Ninth Five Year Plan and Approach to the Tenth Five Year Plan: Draft Document.

Trivandrum Municipal Corporation (2002): Development Seminar: Tenth Five Year Plan (2002-07), Draft Project Document.

United Nations (1962): *Decentralization for National and Local Development,* United Nations, Technical Assistance Programme.

United Nations Development Programme (1998): *Human Development Report*, Oxford University Press, New York, p. 11.

UNDP (2002): Human Development Report 2002, *Deepening Democracy in a Fragmenting World,* Oxford University Press.

Uphoff, Norman (1986): *Local Institutional Development: An Analytical Source Book With Cases,* Kumarian Press, West Hartford, CT.

Uphoff, Norman (1988): "Assisted Self–reliance: Working With, Rather Than For, The Poor", In John P. Lewis [ed] *Strengthening the Poor; What Have We Learned?,* Transaction Books, New Brunswick, NJ.

Uphoff, Norman (1993): "Grassroots Organizations and NGOs in Rural Development: Opportunities with Diminishing States and Expanding Markets", *World Development,* 21 [4], pp. 607–22.

Vijayanand S.M (2001): "Issues Related to Administrative Decentralization and Administering Decentralization: Lessons from the Kerala Experience", paper presented at the Workshop on Decentralization, Institute for Social and Economic Change, Bangalore, 31 May and 1 June, p. 14.

Vijayanand, S.M (2002 a): "Status of Panchayati Raj in Kerala", Paper presented in the IMG Seminar, Thiruvananthapuram, Kerala.

Vijayanand, S.M (2002 b): "Fiscal Decentralization in Kerala: Points for Discussion", Paper presented in the IMG Seminar, Thiruvananthapuram, Kerala.

Visaria, P.D., Gumber, A and Visaria, L (1993): "Literacy and Primary Education in India: 1980-81 to 1990-91, Differentials and Determinants", *Journal of Educational Planning and Administration,* Vol. 7, No. 1, New Delhi.

Voluntary Action Network India [VANI] (1995): *State Panchayat Acts–A Critical Review,* VANI, New Delhi.

Wadia, F.K (1996): "Panchayats in India–Approaches and Ground Realities in a Historical Perspective", *Journal of Indian School of Political Economy,* III January–March, pp. 64–97.

Webster, Neil (1990): "Panchayat Raj and the Decentralization of Development Planning in West Bengal–A Case Study", *CDR Project Paper* 90.7, Centre for Development Research, Denmark.

White, Roland and Smoke, Paul (2005): "East Asia Decentralizes", in *East Asia Decentralizes: Making Local Government Work,* The World Bank, Washington D. C, pp. 1-23.

Winkler, Donald, R (1989): "Decentralisation in Education: An Economic Perspective", Policy Planning and Research *Working Papers*, Education and Employment, Population and Human Resources Department, The World Bank, WPS, 143.

Witte, John F (1990): "Choice and Control: An Analytical Overview", in W. Clune and J. Witte (eds) The theory of *Choice and Control in American Education*, vol. I, New York: Falmer Press.

Wolf, Thomas and Gurgen, Emine (2000): "Improving Governance and Fighting Corruption in the Baltic and CIS Countries: The Role of the IMF", *Economic Issues–21*, International Monetary Fund, Washington D.C, p. 2,7.

World Bank (1987): *World Bank Experience with Rural Development*, Operations Evaluation Department, Report No. 6883, Washington D.C: World Bank.

World Bank (1990): *Peru: Sector Reform and Investment Review*, Technical Department, Latin America and the Caribbean, The World Bank, Washington D.C.

World Bank (2000): *Rural Decentralization in India*, Part III, The World Bank, Washington D.C, pp. 105–136.

INDEX